GEOGRAPHY AND POLITICS IN CENTRAL INDIA

Geography and Politics in Central India

(A Case Study of Erstwhile Indore State)

Ravindra Pratap Singh

CONCEPT PUBLISHING COMPANY NEW DELHI-110015

ISBN 81-7022-025-4

First Published 1987

Published by
Naurang Rai
Conceept Publishing Company
H-13, Bali Nagar
NEW DELHI 110015 (India)

Printed at
New Gian Offset Printers
495 Sahzada Bagh
DELHI - 110035

Dedicated to
My Mother
Shrimati Jagga Devi

Preface

Political geography studies the spatial differences and similarities, which are political in character. It describes political phenomena in their association with other features of the earth's surface, such as landforms, climate, waterbodies and economic activities, or social patterns, which help man in organising the space for political purposes. The present study is a politicogeographic appraisal of physical and cultural landscape of a fascinating region lying on the Central Highlands of India. It aims to explore the impact of geography on the geopolitical evolution and structure, and the resultant political behaviour of the erstwhile Indore State, a premier princely state in the Central India in pre-independence days, which merged with the Indian Union in 1948.

Political area, or space, 'the kernel of political geography', is multi-dimensional in nature. Space is horizontal as viewed from the standpoint of the shape, size, location and natural resources of a political unit. Space is vertical as viewed from the interplay of man's objectives, laws, and economic and cultural tools upon the horizontal plane. Space is temporal as it considers the interaction of horizontal and vertical space at any given period. Space dimensions vary according to the point of projection or point of view, that is applied to them in analysis of spatial relations

Various approaches, such as power analysis, historical, morphologic and functional, have been employed in politico-geographic studies. Power analysis refers to "an analysis of political units of power and the relations among them." Historical political geography has, as its focus, the political landscapes of the past, both for sake of understanding the past better and for analysing current problems. The morphologic approach examines political areas according to their patterns and structural features. Functional political geography is concerned with the functioning of an area as a political unit. The book attempts to study the complex political pattern of a typical Central Indian State in the historical perspective. The functional and morphologic approaches have been blended together the present study to analyse the spatial relationships of an erstwhile political unit. Of course, in such a study it is extremely difficult to avoid the imprints of cultural phenomena. As such, the readers may find in the book a regional cultural geography of Central India with political overtones.

The book is based primarily on my Ph.D. dissertation "Political Geography of Erstwhile Indore State", submitted to the Vikram University, Ujjain (M.P.) in 1968-69 under the supervision of Professor K.N.Verma, the then Head of the Geography Department, Hamidia College of Arts and Commerce, Bhopal.

I am indebted to Maharajkumar Dr. Raghubir Sinhji of Sitamau (Malwa) for his help and advice from time to time; to Professor K.N. Varma for academic supervision of the work to Professor S.P. Chatterjee, formerly Head of the Geography Department, Calcutta University and Professor R.L. Dwivedi, Head of the Geography Department, Allahabad University, who encouraged and inspired me to work in the field of political geography. I am grateful to Professsor L.C. Dhariwal and Shri Surendranath Dube, who held important offices in the Holkar State Administration, for their useful narratives and advice; to Principal Sadashiv Onkarnath Dube, Sailana (M.P.), where I taught for about two years (1962-64); Shri Kanhaiya Lal Dwivedi, Manasa (M.P.); Professor B.M. Singhal, Madhav Mahavidyalaya, Ujjain; Shri M.L.Solanki, Bhopal; and Dr. Gajadhar Singh, formerly Reader in Agriculture, Regional College of Education, Bhopal for their help in many ways in completing this work.

I gladly acknowledge the help of my former student R.K. Samarjit Singh, Cartographger in the University, in drawing of maps for the book.

I sincerely acknowledge my indebtedness to the authors of various books and articles listed in the Bibliography, whose works have immensely helped me in preparation of this book.

Rainder Pratap Singh
Department of Earth Sciences,
Manipur University,
IMPHAL

Contents

List of Maps

List of Tables

CHAPTER 1

Introduction

The Central India - 'roughly the triangle between Delhi, the Gulf of Cambay, and Bengal - is an area of great physical complexity'[1] and immense geopolitical significance. The hill ranges, scraped plateaus and troughs, which form the region, represent the buckling of the northern foreland of the Indian Peninsula under the stress of the Himalayan mountain building. The Central Highlands[2], as this region is sometimes referred to, also include the peneplains of Chota Nagpur, the complex topography of the Maikal block, the highly dissected gneissic terrain of Bundelkhand, and the great northern salient of Deccan lava in Malwa.

The region, forming core of the country, has rightly been called the Madhya Desa or the heart of India. It is an area transitional between the Indo-Gangetic Plains and the area transitional between the passageway between the north and Deccan, and has formed the passageway between the north and south India. Peopled by the primitive tribes like Gonds, Bhils and Santhals with pockets of martial communities like Rajputs and Marathas, who ruled the region prior to Independence, Central India exhibits a unique demographic pattern of its own.

Geopolitically the region formed a great shatterzone[3], accommodating the great majority of the nearly 600 princely states, most of them small and very small both in area and population. Majority of these erstwhile states now form part of Madhya Pradesh and Rajasthan, and a feww with Gujarat. The present day inter-state boundaries in the Central India, with a large number of twists, bends nd curves, reflect the merchland position of the region between the powers of the Indo-Gangetic Plains and the Deccan.

The entire region, particularly Malwa, by its geographical position has been exposed to the cultural impacts from the north well as from the south. It underwent a change of masters more often than any other part of India, and was invariably an apange to the domains of every monarch who became the masters of the Gangetic Plain. Both, during the time of the Mughals, and of the Marathas, Malwa was used as the highway to the south[4].

After the death of the last of the great Mughals, Aurangzeb in 1707, Mughal rule ceased to exist as an effective force in India. None of his successors proved men of ability. Under their feeble rule the disintegration of the Mughal empire was rapid.

Malwa, the richest of the Mughal provinces, was left ungoverned after the death of the Nagar Brothers in the Battle of Amjhea (1728)[5], in which Mughal army recorded its first setback in conflict with the Marathas. The young Malhar Rao Holkar, who showed a high standard of military skill and bravery in the battle, was grated 12 districts in Malwa by Peshwa Baji Rao, which were subsequently increased to 82 in 1731, in return for further valuable services rendered to the Maaratha cause. He was granted 35 per cent share in Malwa province, where is declared the suzerainty of the Peshwa in 1731. Malhar Rao collected taxes and tributes from almost all the kings in Rajputana and Malwa. After taking part in not less than 40 battles, he died in 1766, and left behind a vast state (yielding about a crore an year) enjoying peace and good government[6].

Malhar Rao was succeeded by his grandson, Male Rao, who could survive for more than nine months, and the management of the state devolved solely on his mother Ahilyabai. She succeeded in gaining the affection of her own subjects as well as the confidence of the rulers of the nieghbouring states. She was assisted in her administrative duties by Tukoji Rao I. She was 'one of the purest and most exemplary rulers' who ever lived[7]. She died in 1795 at the age of 70 years, and was succeeded by Tukoji Rao I. He was a General of great tact and ability. He introduced the European system of welfare and discipline in his army. With his death in 1797 ends the first period of Indore's history, which can rightly be called the period of evolution of this state.

A period of conflicts follows the period of evolution in the historical development of Indore State. Two legitimate sons of Tukoji Rao - Kashi Rao and Malhar Rao, could not patch up their differences over the issue of succession. A reconciliation[8] wa effected between two brothers, but powerful Scindia attacked Malhar Rao and killed him in battle-field. Tukoji Rao had two illegitimate sons-Yashwant Rao and Vithoji. After Malhar Rao's death Yashwant Rao rose to power from initial nothingness, entirely by dint of his personal valour and spirit of adventure. He defeated Colonel Monson on the banks of River Banas, near Kushal Gaarh, on 24th August 1804 in course of the Third Anglo-Maratha War. 'No disaster of a similar nature had yet occured to our arms in India'[9], wrote a British historian. He was the last Maratha general to make name in the history. Yashwant Rao died in 1811 at Bhanpura after a prolonged insanity. His four year old son, Malhar Rao II, succeeded him. Tulsi Bai, the most favourite of Yashwant Rao's mistresses, acted as Regent on behalf of the young king. Utter confusion reigned in various parts of the state, and she was compelled to retire in 1815. The Battle of Mahidpur, in December 1817, proved disastrous to the Holkar State. The Treaty of Mandsaur (Appendix III) was concluded in January 6th 1818 with the British Government. It marks the end of the period of conflicts in the history of the Indore State.

An era of peace and progress dawned upon the State after the Treaty of Mandsaur (1818). Indore was made State's capital in place of Maheshwar. The State revenue rose from Rs.5 lakhs in 1817 to 27 lakhs in 1826, under Tantia Jog's ministership. After Malhar Rao (1834-43) and Khande Rao (1843-44) became rulers successfully of the Indore State. Tukoji Rao II succeeded Khande Rao in 1844. The first War of India's Independence broke out in 1857. Between 1861 and 1868, various exchanges of territory were affected to make the State more compact. The Holkar State Railway was constructed in 1870. Maharaja Shivaji Rao succeeded Tukoji Rao II in 1886. Maharaja Shivaji Rao succeeded Tukoji Rao II in 1886. He abolished all the transit duties throughout the State. He abdicted in favour of his son Tukoji Rao III in 1903. At the outbreak of the World war, I, Maharaja Holkar placed all his resources at the disposal of the British Government. Tukoji Rao III too abdicted in 1926 in favour of his son, Yashwant Rao II, who ruled the State till its merger in the Indian Union in 1948.

THE STUDY AREA

Indore State was a premier state in the Central India. Its territories now form part of the states[10] of Madhya Pradesh and Rajasthan (Figure 1). It derived its name from the capital city of Indore. This city was also the headquarters of the Central India Agency, where Agent to the Governor-General in Central India resided. The old city called 'Juni Indore' was a small village in the vicinity of Indreshwar temple, the site of which pleased Ahilyabai, who encamped there after the death of her father-in-law, Malhar Rao Holkar, the founder of the Holkar State. She ordered to shift the headquarters of the district to this village from Kampel. A new town flourished on the

MAP 1

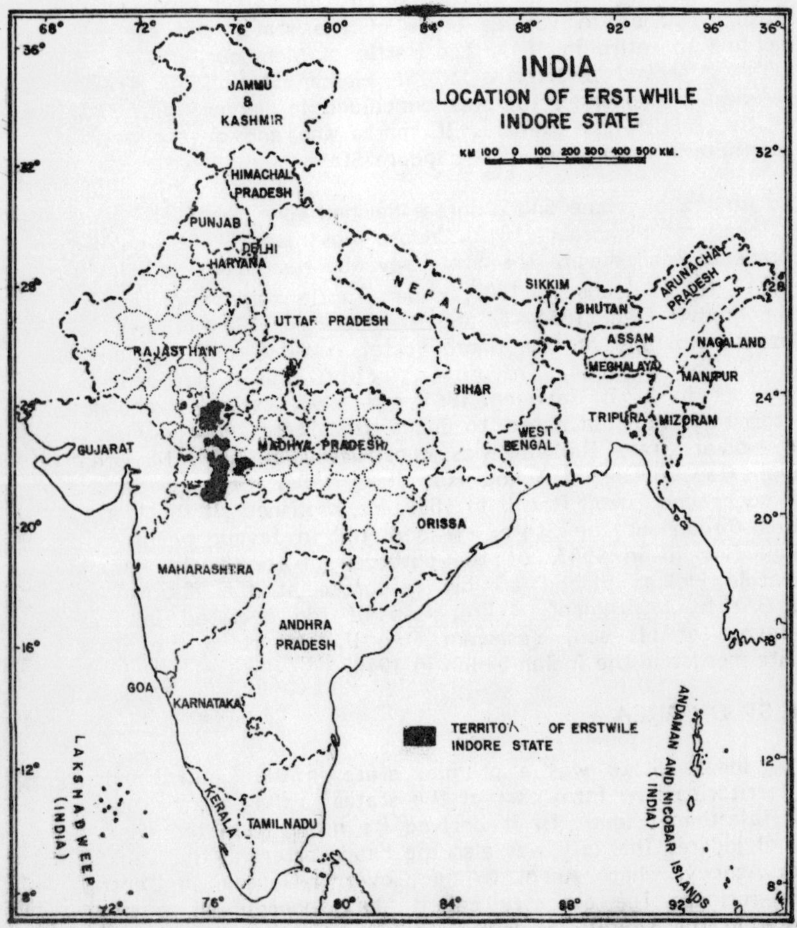
INDIA
LOCATION OF ERSTWHILE
INDORE STATE
KM 100 0 100 200 300 400 500 KM.
JAMMU
&
KASHMIR
HIMACHAL
PRADESH
PUNJAB
DELHI
HARYANA
RAJASTHAN
UTTAR PRADESH
NEPAL
SIKKIM
BHUTAN
ARUNACHAL
PRADESH
ASSAM
NAGALAND
MEGHALAYA
MANIPUR
BIHAR
TRIPURA
MIZORAM
GUJARAT
MADHYA PRADESH
WEST
BENGAL
ORISSA
MAHARASHTRA
ANDHRA
PRADESH
GOA
KARNATAKA
KERALA
TAMILNADU
LAKSHADWEEP
(INDIA)
ANDAMAN AND NICOBAR ISLANDS
(INDIA)
TERRITORY OF ERSTWILE
INDORE STATE

Table 1.1 : Administrative Units of Indore State

Division	District	Mahal	Location in present district*
A. Malwa	I. Indore	1. Indore	Indore
		2. Depalpur	Indore
		3. Sawer	Indore
		4. Mhow	Indore
	II. Mahidpur	1. Mahidpur	Ujjain
		2. Tarana	Ujjain
		3. Jirapur	Rajgarh
		4. Petlawad	Jhabua
		5. Alampur	Bhind
	III. Ramapura-Bhanpura	1. Nandwai	(Chittaurgarh)
		2. Mansa	Mandsaur
		3. Rampura	Mandsaur
		4. Bhanpura	Mandsaur
		5. Garoth	Mandsaur
		6. Sunel	(Jhalawar)
B. Nimar	IV. Nimar	1. Nisarpur	West Nimar
		2. Maheshwar	West Nimar
		3. Barwaha	West Nimar
		4. Kasrawad	West Nimar
		5. Segaon	West Nimar
		6. Sendhwa	West Nimar
		7. Khargone	West Nimar
		8. Bhikangaon	West Nimar
	. Nemawar	1. Kannod	Dewas
		2. Khategaon	Dewas

*All the districts are located in Madhya Pradesh except those in brackets. which are in Rajasthan.

Source : Census of Central India, 1941 Vol. 1 - Holkar State- Tables, p.2.

western bank of River Khan, a small clear stream which flowed past it, with the old name Indore. Ahilyabai's partially fascination for this spot soon raised it to a state of comparative prosperity though she continued to reside at Maheshwar.

The State was divided into two divisions, viz. the Malwa and Nimar Divisions, which were administered by the Divisional Commissioners. The Malwa Division comprised the districts of Indore, Mahidpur (now a part of Ujjain district). The Nimar

Division consisted of the Nimar (now West Nimar district) and Nemawar (now part of Dewas district) districts. These five districts of the state were further divided into 25 mahals[11] as shown in Table 1.1.

There was an uneven distribution of _mahals_ in various districts of the Indore State (Figure 2). Nimar had eight _mahals_ and Nemawar only two. The average area of a _mahal_ was 1,029 km^2, but the range of size was very great. The largest among the _mahals_ was Khargone in Nimar district (3,134 km^2), which was bigger than the entire district of Nemawar (2,792 km^2). The smallest _mahal_ was Nandwai in Rampura-Bhanpura district (113 km^2), followed by Alampur in Mahidpur district (114 km^2). In general the area of districts and _mahals_ had a very close relationship to the value of land and the density of population (Table 1.2). In areal extent the mahals were large enough in the tribal belt of the State, and too small in the low-lying densely populated northern enclaves. Most of the _mahals_ (21 out of 25) had an area of less than 1,250 km^2 and a population of below one lakh.

These _mahals_ of the Indore State were organized merely for administrative convenience. Their boundaries were always open to adjustments, and therefore, their numbers too were fluctuating from time to time.

The Indore State comprised seven districts and 53 _mahals_ in 1901. But there were certain changes soon after, resulting in reduction of number of the _mahals_ to 52 in 1903. in 1904, the number of districts was reduced to six, by combining Mandaleshwar and Khargone into one district, which was named Nimar. The number of _mahals_ ws also reduced to 37. In 1908, the number of districts was further reduced to five, by combining Ram pura and Bhanpura into one, and the number of _mahals_ was fixed at 27. This arrangement continued till 1929, when Sundarsi was abolished as a separate _mahal_ and was merged into Tarana _mahal_, reducing the number of _mahals_ to 26.[12] After 1931 only one change was affected in the arrangement of _mahals_, which was the abolition of Kantaphod _mahal_ of the Nemawar district, and its amalgamation with Kannod _mahal_, while the eastern part of the latter _mahal_ was merged with _Khategaon_ mahal.[13] This arrangement continued till the State was merged to form the erstwhile Madhya Bharat Union, now part of Madhya Pradesh.

The ruler of Indore State was very progressive in outlook. He introduced democratic institutions in the state in the year

Table 1.2 : Area and Population of Administrative Units

State/District/Mahal	Area (km^2)	Population (1941)
INDORE STATE	25,729	15,13,966
Indore District	3,820	4,50,921
Indore	1,015	2,62,650
Depalpur	1,031	60,754
Sawar	694	48,878
Mhow	1,080	81,819
Mahidpur District	3,590	2,31,022
Mahidpur	1,124	67,330
Tarana	1,1[illegible]9	74,998
Jirapur	8[illegible]3	51,738
Petlawad	4[illegible]0	20,289
Alampur	[illegible]14	16,858
Rampura-Bhanpura District	4,722	2,08.752
Nandwai	113	5,205
Manasa	941	57,287
Rampura	[illegible],215	32,352
Bhanpura	922	26,696
Garoth	1,142	60,508
Sunnel	389	26,704
Nimar District	10,805	5,25,199
Nisarpur	1,176	57,847
Maheshwar	725	51,766
Barwaha	1,140	62,278
Kasrawad	818	46,483
Segaon	1,134	67,530
Sendhwa	1.062	68,398
Khargone	3.134	96,868
Bhikangaon	1,616	64,029
Memwar District	2,792	98,072
Kannod	1,681	55,367
Khategaon	1,111	42,705

Source : Census of Central India, 1941 Vol. 1 - Holkar State - Tables, p.2.

1909, when popular government was not established in any part of the country. For the popular representation in the governmental affairs, the State organized and constituted local boards at various levels. The city of Indore and other towns were provided with municipalities, while for other areas *gaothi* (village) panchayats, *mahal* boards and district boards were constituted.[14] This pattern is now being followed throughout

MAP 2

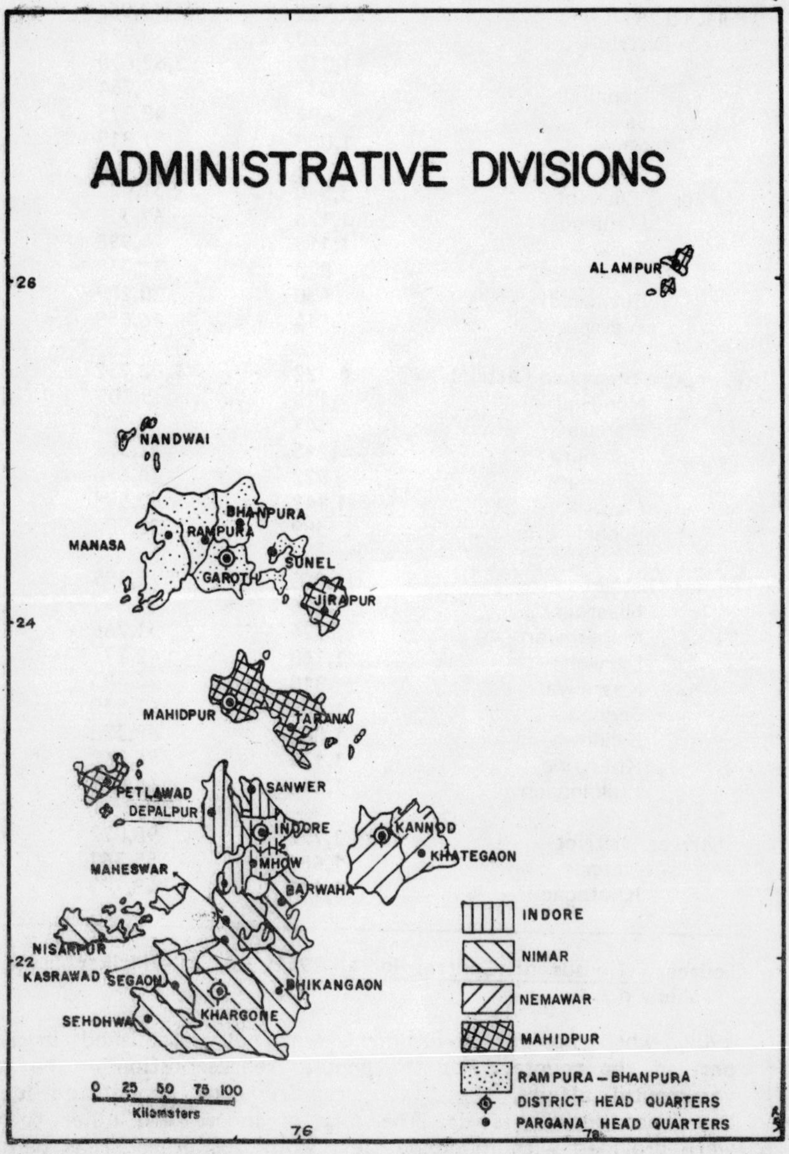
ADMINISTRATIVE DIVISIONS
26
24
22
76
78
ALAMPUR
NANDWAI
BHANPURA
RAMPURA
MANASA
GAROTH
SUNEL
JIRAPUR
MAHIDPUR
TARANA
PETLAWAD
DEPALPUR
SANWER
INDORE
MHOW
KANNOD
KHATEGAON
MAHESWAR
BARWAHA
NISARPUR
KASRAWAD
SEGAON
BHIKANGAON
KHARGONE
SEHDHWA
INDORE
NIMAR
NEMAWAR
MAHIDPUR
RAMPURA – BHANPURA
DISTRICT HEAD QUARTERS
PARGANA HEAD QUARTERS
0 25 50 75 100
Kilometers

the country. Through these boards the people of the state participated in the administration of their local affairs, especially the public utility services viz. water supply, sanitation and health services, lighting, communication, transport etc.

OBJECTIVES

The state has been accepted by geographers generally[15] as the formal or central subject matter of political geography. Its political significance bears a well defined relation to its climate, landforms and natural resources.[16] The present study attempts a politico-geographic appraisal of physical and cultural landscape of a fascinating region lying on the Central Highlands of India. The study aims to explore the impact of geography on the geopolitical evolution and structure, and the resultant political behaviour of the erstwhile Indore State, which merged with the Indian Union in 1948.

METHODOLOGY

The observational and survey research methods have been adopted in conducting this study. It is based on published materials - books, journals and government reports. Field observation and spot studies were carried on for checking up of the available data. The officials of the erstwhile Indore State were interviewed to find out the people's participation in the state-craft. The collected information has been organized, mapped, described and analysed to explain the nature of the present study.

A SHORT PLAN OF THE WORK

The first chapter, including this section, introduces the basic politico-geographic characteristics of the Central India, of which the erstwhile Indore state territories have been the core, and discusses the objectives and methodology of the study. The second chapter deals with the locational perspective and its impact on the political behaviour of the State. The geopolitical evolution of the former Indore State has been attempted in the third chapter to emphasize the influence of historical forces on her existence as a premier political unit in the Central India in pre-Independence days.

Chapter 4 deals with the ecological setting of the State. It portrays briefly the various environmental factors, viz. relief, climate, soils and natural vegetation, which are considered to be the integral parts of the physical and political landscapes. This gives an understanding of their limitations in relation to land and its human use. In chapter 5, the economic resources of the State are critically reviewed. Economic and political activities in any political area are intricately intertwined. States, therefore, rely upon economic strength for their political power. Hence, an analysis of agriculture and associated activities, minerals and power, industries and trade of the State have been made in this chapter.

The circulatory system of transportation and communication networks have been discussed in Chapter 6 because 'it is essential for the political organization of space and to the maintenance of this organization'.[17]

Man has not only been an asset on the earth's surface but also a liability. He is the creator and designer of the whole cultural complex, which can be treated as the laboratory of the geopolitics. As such in Chapter 7, an attempt has been made to discuss the geopolitical aspect of the demographic pattern in the State.

Chapter 8 depicts the social structure and its impact on unity and integrity of the people in this erstwhwile State. Besides the caste system prevalent in this region, the customs and political importance.

The lst chapter sums up the findings and draws the broad conclusions.

FOOTNOTES

1. O.H.K. Spate and A.T.A. Learmonth, India and Pakistan - A General and Regional Geography, London, 1967, p. 611.
2. The Gazetteer of India, Vol. I, Delhi, 1965, p. 40.
3. A 'shatterzone' is a large strategically located region, which is occupied by a number of conflicting states, and is caught between the conflicting interests of great powers which adjoin or have access to it.
4. V.P. Menon, The Story of the Integration of the Indian States, Calcutta, 1956, p. 223.
5. G.S. Sardesai, Selections from Peshwa Daftar, Part XIII, Letter Nos.23, 25, 27 ; Raghubir Sinh, Malwa in Transition - A Century of Anarchy, Bombay, 1936, p. 164.
6. S. Playne, Indian States - A Biographical, Historical and Administrative Survey, London, 1921, p. 76.
7. John Malcolm, Memoir of Central India, Vol. I, New Delhi, 1970, p. 194.
8. 'Scindia' is also referred to as 'Sindhia'.

9. G. Mac Munn, Indian States and Princes, London, 1936. p.69.
10. 'States' in the present politico-administrative set-up of India signify 'provinces'. The Constitution of India, Delhi. 1950, Article 1.
11. 'Mahal' signifies a sub-division, tahsil or pargana.
12. L.C. Dhariwal, The Indore State Gazetteer, Vol. I, Indore, 1931, p. 412.
13. Surendranath Dube, Census of Central India, Vol. I- Holkar State - Tables, Indore, 1941, p.1.
14. Dhariwal, op. cit., p.339.
15. W.A.D. Jackson, "Whither Political Geography", Annals of the Association of American Geographers, Vol. 48, 1958, p. 178.
16. Derwent Whittlesey, The Earth and the State, New York, 1944, p. 585.
17. Lewis M. Alexander, World Political Patterns, London, 1957, p. 41.

CHAPTER 2

Locational Perspective

The erstwhile Indore State had acquired its popular name from its capital city Indore, more correctly Indur, which appears to be a derivative of Indrapur,[1] the present city has grown. The State comprised the former dominions of His Highness the Maharaja Holkar, and therefore, sometimes was also referred to as the Holkar State.

LOCATION

The State was situated principally in the Central India tracts of Malwa and Himar. It consisted of severl large and small blocks of territory, lying between 21°22' and 24°40' worth latitude and 74°22' and 77°30' east longitude. But when the isolated pargana (mahal) of Nandwai, lying in the midst of erstwhile Udaipur State, and the detached pargana of Alampur, surrounded by the territories of former states of Gwalior and Datia, were also included, the ultimate limits of the State extended to 21°22' and 26°6' north latitude and 74°22' and 26°6' north latitude and 74°22' and 78°51' east longitude.[2] The Tropic of Cancer, thus, divides the State into two halves-northern and southern.

The State was not very far from the sea board. The open Arabian Sea was only about 250 to 300 km from her southern part, and was well connected by roads and rail routes.

The State had very complex civinal location (Figure 3) as it was surrounded by a number of princely states of the then Rajaputana, Bundelkhand, Bhopal and Central India Agencies. These erstwhile states numbered 18, i.e. Datia of Bundelkhand Agency; Udaipur, Kota, Pratapgarh, Jhalawar and Kushalgarh of Rajputana Agency; Gwalior, Jaora, Tonk, Dewas, Dhar, Ratlam, Sailana, Jhabua, Alirajpur and Barwani of Central India Agency; and Khilchipur and Bhopal Agency. The southern boundary of the State was formed by the then British Indian districts of East and West Khandesh, Nimar (Khandwa) and Hoshangabad.

MAP 3

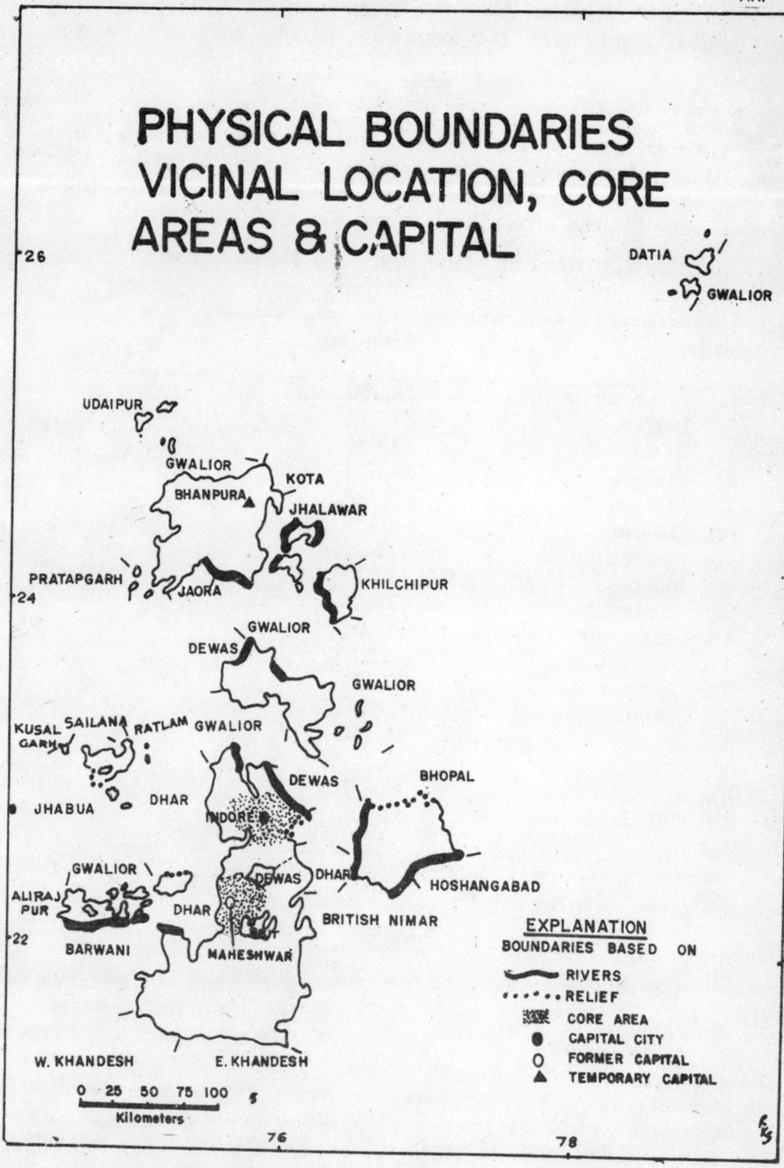
PHYSICAL BOUNDARIES
VICINAL LOCATION, CORE
AREAS & CAPITAL
26
DATIA
GWALIOR
UDAIPUR
GWALIOR
KOTA
BHANPURA
JHALAWAR
PRATAPGARH
JAORA
KHILCHIPUR
24
GWALIOR
DEWAS
GWALIOR
KUSAL GARH
SAILANA
RATLAM
GWALIOR
DEWAS
BHOPAL
JHABUA
DHAR
INDORE
GWALIOR
DEWAS
DHAR
ALIRAJ PUR
DHAR
HOSHANGABAD
BRITISH NIMAR
22
BARWANI
MAHESHWAR
EXPLANATION
BOUNDARIES BASED ON
RIVERS
RELIEF
CORE AREA
CAPITAL CITY
FORMER CAPITAL
TEMPORARY CAPITAL
W. KHANDESH
E. KHANDESH
0 25 50 75 100
Kilometers
76
78

Strategically her location was of prime importance having a base on the marchland between the powers based in the Indo-Gangetic Plain and those of the Deccan. It was all due to her route-land situation that the region saw a constant change of masters, right from the beginning of her recorded history.

SIZE

With an area of 25,729 km^2, the State wss of very small size. Among the then princely states of India it ranked 15th in area. In the Central India Agency, however, it was the third largest state in area after Gwalior and Rewa (Table 2.1).

Table 2.1: Area of Major Erstwhile Indian States

States	Area (km^2)
1. Bahawalpur	38,850
2. Baroda	21,070
3. Bastar	33,831
4. Bhutan	46,620
5. Bikaner	60,386
6. Gwalior	68,332
7. Hyderabad	2,34,188
8. Indore	25,729
9. Jaipur	43,206
10. Jaisalmer	41,601
11. Jaammu and Kashmir	2,22,442
12. Jodhpur	90,821
13. Kalat	1,89,790
14. Mysore	76,340
15. Patiala	15,34
16. Rewa	33,670
17. Udaipur	33,450

Source : G. Mac Munn. Indian States and Princes, London, 1936, pp.266-74.

SHAPE

The State was extremely incompact, and was composed of a large number of isolated tracts of territory, scattered about in the Central India, but chiefly situated on the Malwa plateau and in the Narmada valley. To enumerate the blocks of territory of the State, starting from the north, the first to be encountered was the Alampur pargana, which was composed of five small isolated tracts of the Holkar territory, surrounded by the Old states of Gwalior and Datia, and situated in the Bundelkhand upland, about 80 km to the North-North-East of Jhansi. Descending southwards, about 398 km to the south-west of Alampur was the main block of Rampura-Bhanpura district on the Malwa plateau. To the north-west of this block was the Nandwai enclave in former Udaipur state. To the east were the parganas of Sunel and Jirapur. Proceeding southwards for about 32 km, there was the main block of the Mahidpur district. The Indore district was again 32 km south of Mahidpur district. The Petlawad pargana of Mahidpur district was about 64 km west of Indore district, while the Nemawar district in the Narmada basin was about 32 km in the east of it.

Passing from the Malwa plateau to the Narmada valley, to the south of the Indore district, lay the district of Nimar. It has a panoramic situation between the Vindhya and Satpura ranges, and traversed by the river Narmada. The Nisarpur pargana of this district was about 24 km west of the main block.[5]

Thus the state had a fragmented shape as there were nearly 80 disconnected units, every one of which was entirely surrounded by the territories of other states. The only tie, uniting these divergent component parts of the State, was the common administration.

FRAGMENTATION OF STATE'S TERRITORIES

The scattered nature of the State territories furnished some explanation for the rise and fall of the Maratha power, as well as the brilliant rise and subsequent diminution in the fortunes of the House of Holkar. The Maratha conquest was carried into Malwa by Holkar, Scindia and Pawar. These Maratha generals penetrated even into Rajputana and north Indian plains, and carved out large kingdoms for themselves. The territories of Holkar, Scindia and Pawar were interlaced as after every conquest the region used to be divided among the three. Thus, in the Narmada valley, the Nimar and Nemawar districts of Holkar adjoined Nimanpur, Thikri, Dharampuri, Kukshi and Bagod parganas of Pawar ; and Bankaner, Manpur and portions of the Nemawar district of Scindia, of which Scindia retained only Bakaner. Others, viz. Manpur and portions of Nemawar came to Holkar through the British, to whom Scindia had assigned these tracts.

In western Malwa, the Badnawar pargana of Pawar, the Amjhera district of Scindia and the Petlawad pargana of Holkar wer intermingled with each other. In southern Malwa, Holkar's Indore district stood close to the Ujjain district of Scindia, and the Dewas and Dhar parganas of Pawar. In the extreme north of Malwa and adjoining areas of Rajputana, the Mandsaur district of Scindia had its counterpart in the Rampura-Bhanpura district of Holkar. In the central part, the Mahidpur district of Holkar[1] stretched along the Shajapur district of Scindia, and Allot pargana of Pawar bordering it in the west.

Later on when the Holkar and the Scindia carried independent banners, their common conquests ceased and consequently no occasion arose for a division of the conquered territory. The territories, independently acquired by Holkar, were mostly lost in the last struggle against the British. The military successes of Maharaja Yashwant Rao Holkar I roused the apprehensions of the British, and at the end of his rule the Holkars lost much of their possessions. The struggle, which ended with the battle of Mahidpur in 1818, reduced the extent of the Indore State, which remained as such till its merger in the Indian Union in April 1948, except slight modifications by later territorial exchanges.

CAPITAL

Till 1818, the capital of the Holkar State was Maheshwar on river Narmada, where the rulers used to live. Maheshwar is the Mahishmati or Mahissati of early days. Katyayana in his commentry on Panini states that Mahishmati derives its name from the prevalence of buffaloes (mahisha) in that region. It is connected with the Pandava brothers, and is mentioned in the Ramayana and Mahabharata, while the Puranas refer to the Mahishakas as the people of Mahishmati.[6] It formed part of the kingdoms of the Haihaya chiefs, the Chalukyas, the Farmers, and of Ahmad Shah of Gujarat. It formed part of the Mughal empire and later on in about 1730, it passed into the possession of Malhar Rao Holkar I, the founder of the Holkar State.[7]

After the death of Malhar Rao I in 1766, the administration of the State fell on Male Rao, his grandson, who died within nine months of his succession. His mother, Ahilyabai, took over the charge of the State, and lived at Maheshwar, but her Commander-in-chief, Tukoji Rao I, lived at Indore. Thus the Holkar State those days had two capitals - Maheshwar and Indore. Indore was purely a military headquarter. Yashwant Rao Holkar, son of Tukoji Rao, directed the affairs of the State from Bhanpura, but Malhar Rao II shifted his capital to Indore in 1818. Indore remained the capital of the Holkar State till 1948, when after the State's merger in the Indian Union, it became the second capital of the then Madhya Bharat as per Nehru Award.[8]

Indore as the capital city of the State was centrally located, and was linked by roads and rail-routes with every important town of the State. Only Alampur pargana was slightly away from Indore city, which remained in the State due to historical necessities.[9]

References

1. Malav Itihas Mandal, The Mandalik Papers ad the Family, Indore, 1946, p. 10; L.C. Dhariwal, The Indore State Gazetteer, Vol. 1, Indore, 1931, p. 1. The Imperial Gazetteer of India, Vol. XIII, Oxford, 1908, p. 333 cites the name of the village as 'Indreshwar', and not Indrapur.

2. Holkar State Administrative Report 1945, Indore, 1947, p.1. These figures slightly differ from those given in the Imperial Gazetteer of India, Vol. XIII, where it is mentioned as 21°22' and 26°3' north latitude and 74°30' and 78°51' east longitude.

3. The number of states have been calculated on the basis of the map attached to the Census of Central India 1941, Vol. 1 - Holkar State.

4. Surendranath Dube, Census of Central India 1941, Vol. 1 - Holkar State - Tables, Indore, 1941, pp. 1-2.

5. The distances between various blocks of the /state territory have been taken from the Census of Central India 1931, Vol. XVI - Holkar State, Part I-Text, Indore, 1931, p.1.

6. Dhariwal, op. cit., p. 623.

7. John Malcolm, Memoir of Central India, Vol. 1, London, 1823, p. 121.

8. V.P. Menon, The Story of the Integration of the Indian States, Calcutta, 1956, p. 238.

9. Malhar Rao I died at Alampur on May 20, 1766, where a Chhatri (cenotaph) stands in his memory. Dhariwal, op. cit., p. 24.

CHAPTER 3

Geopolitical Evolution

Malwa was one of the most notable provinces of India during the Hindu and early Muslim periods of our history, and its influence on Hindu culture was of an abiding character.[1] Historically it has been a transitional zone between the plains of Northern India and Deccan. The powers based in the Great Indian Plain and in the Deccan had always aspired to keep Malwa under their control.

This Central Indian tract was probably once occupied by the ancestors of the Bhils, Gonds, Sahariyas and other tribes which now inhabit the fastness of the Vindhya range. The early Hindu scriptures - the Rig Veda, Ramayan and Mahabharata, describing the spread of Aryans and their struggle with the dark-skinned aboriginals, assign the hilly Vindhya region and the Narmada valley to the non-Aryan Pulindas and Sabaras.[2]

During Buddha's lifetime Malwa (Avanti), with Ujeni (Ujjain) as its capital, was one of the sixteen principal states of India. It formed part of the empires of the Mauryas, the Sakas, the Guptas, Harshvardhana and Bhoj. When the Rajputs established their sway over the greater part of the northern India, Malwa also came under their control and Rajput kingdoms flourished there. These kingdoms later succumbed to the might of the Ghoris, Khiljis, Sher Shah and Akbar, the Great Mughal. Under the Mughals, Malwa was a compact unit and formed a 'Subah'.[3]

With the death of Auramgzeb in 1707, Mughal rule in India ceased to exist as an effective force. None of his successors proved men of any ability. 'Delhi became a mere cockpit of warring factions and the throne was occupied by a series of <u>rois faineants</u> under whose feeble rule the disintegration of the empire rapidly proceeded.'[4]

Mughal emperor Muhammad Shah, becoming suspicious of Nizam-ul-Mulk of Hyderabad, who was rapidly consolidating his position as an independent ruler, appointed Girdhar Bahadur as Governor of Malwa in 1724. Sawai Jai Singh of Jaipur thought of relief for Hindus from tyranny of the Mughals, but Girdhar

Bahadur proved to be more cruel than the Muslim governors. Raja Jai Singh sent a diplomatic messenger through Rao Nandlal Mandlik of Indore[5] to Peshwa Baji Rao I, who was destined to revolutionize Maratha politics, and invited him to Malwa.[6]

The Peshwa took up the hint. In early October, 1728, the Maratha army advanced from the Deccan under the overall command of Chimnaji, accompanied by Udaji Pawar and Malhar Rao Holkar. The Maratha army reached the southern bank of the Narmada on 24th November, 1728, and after five days on 29th November, it won the grand victory of Amjhera, near Dhar, after a stiff fight in which both Girdhar Bahadur and his cousin, Daya Bahadur, lost their lives.[7] The news of Maratha victory soon spread throughout India and shocked the Mughal court, as much as it gladdened the Maratha hearts.

PERIOD OF EVOLUTION (1730-1797)

For the prominent part played by Malhar Rao Holkar in the Maratha expeditions in Malwa, he received a saranjam (grant) of 11 mahal (parganas) in 1726-27.[8] The Peshwa granted 74 parganas of Malwa to Malhar Rao and issued sanad to its effect on 3rd October, 1730.[9] This grant marks the beginning of the evolution of the Indore State. (viz. Harsol, Sawer, Barlqi, Depalpur, Hatod, Mahidpur, Jagoti, Karanj and Makdon).[11] These Khasgi lands in Malwa, as they were termed, were expectd to yield Rs.2,63,000 a year, a sum which was not to be included in the saranjam of Holkar. This grant evidently marked the beginning of the erstwhile Indore State.

Malhar Rao Holkar ws granted a joint saranjam with Ranoji Scindia of five mahals of Nemawar district, viz. Nemawar, Hirangaon, Kataphod and Satwas, Rajore and Nimanpur on 3rd June, 1745.[12] He received a sum of Rs.64 lakhs and the districts of Rampura-Bhanpura and Tonk in return of his services to Raja Madho Singh of Jaipur.[13]

In 1748, in response for his zeal and gallantry during the whole of campaign against the Rohillas in Avadh, the Mughal Emperor granted him a deed for the Sardeshmukhi (a due of 121/2 per cent on the revenue) of Chandore, which was the only royal grant in possession of the Holkar family.[14]

Due to differences of opinion regarding the techniques of warfare, the Holkar did not participate in Afgan-Maratha war in Panipat in 1761. Many historians have considered this event as Holkar's disloyalty to the Peshwa. 'Malhar Rao Holkar

MAP 4

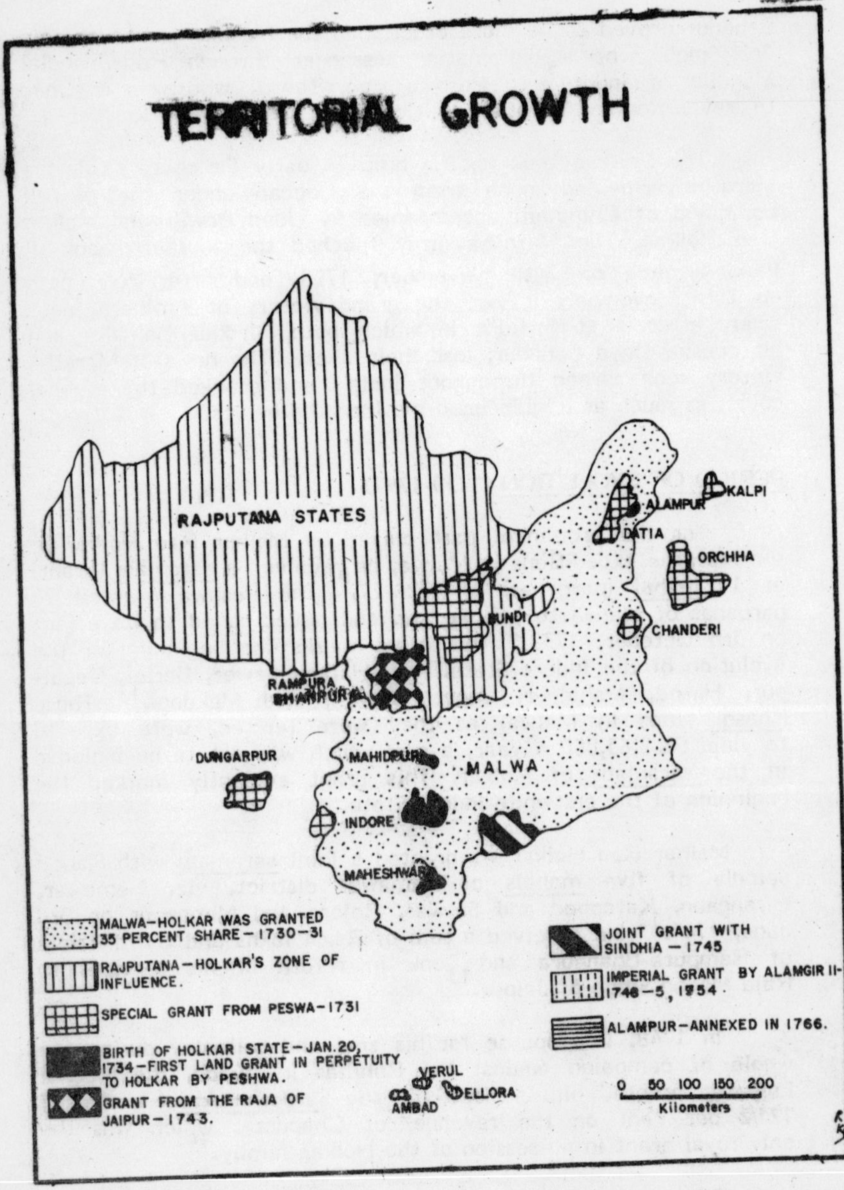
TERRITORIAL GROWTH
RAJPUTANA STATES
KALPI
ALAMPUR
DATIA
ORCHHA
BUNDI
CHANDERI
RAMPURA
BHANPURA
DUNGARPUR
MAHIDPUR
MALWA
INDORE
MAHESHWAR
MALWA—HOLKAR WAS GRANTED 35 PERCENT SHARE—1730-31
RAJPUTANA—HOLKAR'S ZONE OF INFLUENCE.
SPECIAL GRANT FROM PESWA—1731
BIRTH OF HOLKAR STATE—JAN. 20, 1734—FIRST LAND GRANT IN PERPETUITY TO HOLKAR BY PESHWA.
GRANT FROM THE RAJA OF JAIPUR—1743.
VERUL
AMBAD
ELLORA
JOINT GRANT WITH SINDHIA—1745
IMPERIAL GRANT BY ALAMGIR II—1748-5, 1754
ALAMPUR—ANNEXED IN 1766.
0 50 100 150 200
Kilometers

had a genuine difference of opinion on the issue of military tactics. He was not used to artillery warfare, and was out and out an advocate of the guerilla system. But such differences of opinion cannot be construed as disloyalty.'[15]

Malhar Rao, the founder of the Holkar State, breathed his last in April, 1786. He took part in no less than 40 battles. He left behind a vast State, yielding about a crore of rupees per annum, and enjoying the benefits of peace and good government.[16] 'He, in fact, be deemed from the invasion of Malwa till his death (a period of more than 30 years), the most distinguished of the military commanders of the Maratha Empire ; and he appears from his continual employment, to have been greatly favoured by Raji Rao, to whose authority he continued, even in the zenith of his power, to pay the greatest deference.'[17]

Malhar Rao was succeeded by his grandson, Male Rao, who died after nine months. After his death, his mother Ahilyabai took over the charge of administration. To bear titular honours, command her army in the field, and discharge such duties as from her sex she was unable to perform, she selected Tukoji Rao Holkar,[18] a distinguished commander in her army. 'She certainly appears, within her limited sphere, to have been one of the purest and most exemplary rulers that ever existed'.[19] Tukoji shared the power with Devi Ahilyabai. He never forgot, for a moment, his sense of obligation.

Under the administration of Ahilyabai, the territories of the Holkar State remained the same. 'Her toleration, justice and careful management of the State's affairs soon shown in the increased prosperity of her dominions, and the peace which ruled throughout her days.'[20]

On the death of Ahilyabai on the 13th August, 1795, the rule of the State devolved on Tukoji Rao. He followed the example of Devi Ahilyabai, and during his life time the prosperity of the State continued. He died on 15th August, 1797. He participated in many battles and left behind him the character of a good soldier, a plain unaffected man an one, whose courage was superior to his craft.[21] Thus closes the first period of the Holkars' history.

PERIOD OF CONFLICTS (1798-1817)

The death of Tukoji Rao proved disastrous to the interests of the Holkars, as it marked the commencement of a period of prolonged strife between his sons for securing the gaddi (throne) of the Holkars. Tukoji Rao left two legitimate sons

Kashi Rao and Malhar Rao; and two illegitimate sons - Yashwant Rao and Vithoji. Malcolm states that Ahilyabai and Tukoji Rao desired Kashi Rao and Kalhar Rao to occupy positions similar to those occupied by themselves - Kashi Rao to be the administrative head at Maheshwar, and Malhar Rao started contending for the gaddi. Malhar Rao threw himself on the protection of the Peshwa, while Kashi Rao secured the support of Scindia. Malhar Rao was attacked and killed. Yashwant Rao and Vithoji, however, escaped.

Yashwant Rao, after a fugitive life spent partly as a prisoner at Nagpur, and partly at Dhar, managed at length to raise a force and appeared as the champion of Khande Rao, a posthumous son of Malhar Rao.[23] He called on all the old adherents of the House of Holkar to join his standard. He was soon after joined by Wazir Husain of Sarangpur, and later by Amir Khan, whose descendants held the old Tonk State in Rajasthan. He defeated Kashi Rao's troops under Dudrenec at Kasrawad, whereupon Dudrenec transferred his allegiance and his batallions to Yashwant Rao, who entered the capital town of Maheshwar. There 'he ascended the family throne (beginning of January, 1799). All the usual gifts, appointments and rejoicings of a royal coronation now took place.'[24]

Soon afterwards yashwant Rao was defeated at Satwas by some of the Scindia's batllions and retired to Indore, but subsequently attacked Ujjain and extracted a large sum from its inhabitants. 'In October, 1801, Sarje Rao Ghatke, the notorious minister of Daulat Rao Scindi, sacked Indore, practising every kind of atrocity on the inhabitants and razing the town to the ground.'[25] Yashwant Rao, however, assisted by Amir Khan and his Pindaris, started for Deccan, and on 25th October, 1802, he defeated the combined forces of Raji Rao and Scindia, seized the city of Pune. The defeated Peshwa signed the Treaty of Bassein with the British.

The Treaty of Research estranged Scindia, Bhonsle and Holkar from the Peshwa, as they saw in it the disintegration of the Maratha empire. Scindia and Bhonsle entered into a league to fight with the British, and they invited Yashwant Rao to join them. Yashwant Rao could not participate in the conflict between the Scindia and Bhonsle on one hand and the British on the other due to the weak and uncertain economic position of his State. The defeated Scindia and Bhonsle had to sue for peace and this was granted on their making large concessions of territories and rights to the British.

These events raised the position of Yashwant Rao Holkar for higher than that of Scindia and Bhonsle, and the British Government naturally looked upon him as a rival in the contest for supremacy in India. he had also by this time secured from the Delhi Emperor the title of Maharajadhiraj Raj Rajeshwar Alija Bahadur, which gave him a high status amongst the princess of India.[26]

In 1804, after breaking down of the negotiations with the British, Yashwant Rao came into collision with the British forces. In Mukundwara Pass he gained a signal success over Colonel Monson, but was defeated by Lord Lake at Dig on November 13th, 1804.[27]

On 24th December, 1805, he concluded the Treaty of Rajpur Ghat (Appendix 1), on the banks of Beas river, the first engagement entered into between the British Government and the House of Holkar. By this treaty Holkar renounced his claims to Tonk, Bundi and all places north of the Bundi hills and in Baghelkhand; while the British on their part, restored him possessions south of the Chambal and the Deccan with certain exceptions. Lord Cornwallis' policy of non-interference, however, gave him another chance to regain his position. The Rajputana districts were restored to him in February 1806, under the terms of the declaratory article annxed to the Treaty of Rajpur Ghat (vide Appendix I). Thus Holkar succeeded virtually in securing all his possessions and rights, including the right over Jaipur, Udaipur, Kota, Bundi and other Rajput princes.

Yashwant Rao Holkar rose to power from initial nothingness, entirely by dint of his personal valour and spirit of adventure. He was the last typical Maratha soldier to make a name in history.[28] His letter to Vyankoji Bhonsle of Nagpur (Appendix II) gives a vivid picture of the most unfortunate days of our history. So great was his personality that even in those troublesome times (when in this great Holkar insanity developed in his last days), no state or power could venture to commit aggression on his territory; and this influence kept the Holkar State secure even after his death for some years. Yashwant Rao breathed his last on October 28th, 1811 at Bhanpura.

Yashwant Rao left no legitimate heir; but before his death his concumbine, Tulsi Bai, had adopted his illiegitimate son, Malhar Rao, who was placed on the gaddi and whose title was universally acknowledged.[29]

At the outbreak of the war in 1817 between the British and the Peshwa, the Indore State sided with the Peshwa. As the army under Amir Khan, Nawab of Tonk,[30] was about to proceed to Deccan, the news of arrival of Sir John Malcolm at Agar, and Sir Thomas Hislop at Ujjain caused utmost consternation. Malcolm tried for friendly settlement of the terms but military commanders would not come to terms. The battle of Mahidpur was fought on December 21, 1817. After severe fighting the disloyal forces of the Holkar were defeated. Malhar Rao escaped and retreated to Mandsaur. His prime minister, Tantia Jog, negotiated the terms with the British, and the Treaty of Mandsaur. His prime minister, Tantia Jog, negotiated the terms with the British, and the Treaty of Mandsaur (Appendix III) was signed on January 6, 1818 and was ratified by the Governor-General on the 16th January, 1818.

By the Treaty of Mandsaur, the Holkar state was deprived of almost two-thirds of its territories (Figure 5) and rights. Nawab Amir Khan of Tonk was made independent in respect of his holdings. Maharaha ceded to the British Government all claims of tribute and revenue of every description, which he had or may have had upon the Rajput princes, such as the Rajas of Udaipur, Jaipur, Jodhpur, Kota, Bundi etc. He renounced all rights and titles from the territories within or north of the Bundi hills; and ceded to the British Government all his territories within and south of the Satpura hills, posessions in Khandesh and districts such as Amber, Ellora, and others, intermixed with the territories of the Nizam and Peshwa. He retained the tributory rights over the minor princely states - Pratapgarh, Narsingarh and Jhabua. The Maharaha under the provisions of this treaty, granted the paraganas of Sunjeet, Malhargarh, Tal-Mundawal, Jaora, Bardawad (Baroda) with tribute of Piploda to Nawab Gaffoor Khan.[31] The British Government undertook to keep up a field force sufficient to protect the territory from aggression and maintain its tranquility, giving the status of a protectorate to the Holkar State.

PERIOD OF PEACE AND PROSPERITY (1818-1948)

Indore was a state of peace and prosperity under Tantia Jog's prime ministership, who died in 1826. Malhar Rao died on 27th October, 1833, and was succeeded by his adopted son, Martand Rao.[32] Later on, he was replaced by Hari Rao, under whom the administration of the State went bad to worse. His adopted son, Khande Rao, succeeded him in November, 1843, who suddenly died on 17th February, 1844, at the age of 15. Kesri Bai, the venerable widow of Yashwant Rao, nominated second son of Bhau Holkar for the gaddi, who was installed on 27th June, 1844, as Tukoji Rao Holkar II with all the formality of a heriditary chief.

MAP 5

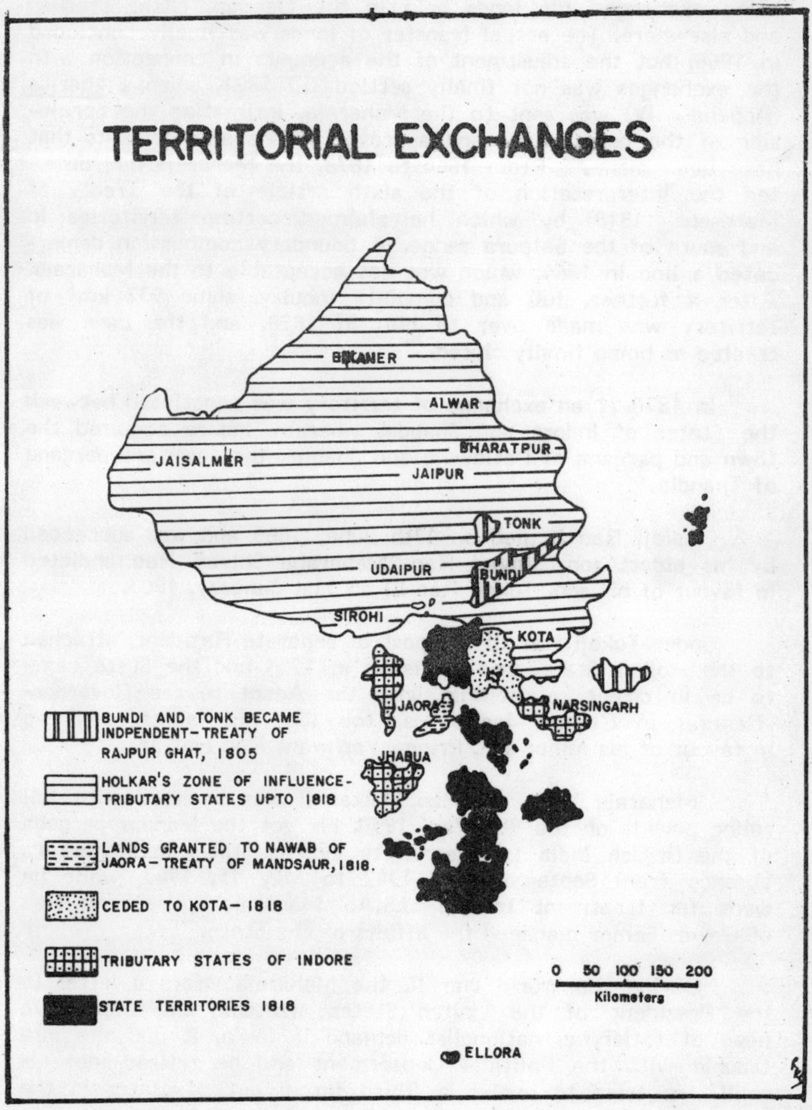
TERRITORIAL EXCHANGES
BIKANER
ALWAR
BHARATPUR
JAISALMER
JAIPUR
TONK
UDAIPUR
BUNDI
SIROHI
KOTA
JAORA
NARSINGARH
JHABUA
ELLORA
BUNDI AND TONK BECAME INDPENDENT-TREATY OF RAJPUR GHAT, 1805
HOLKAR'S ZONE OF INFLUENCE-TRIBUTARY STATES UPTO 1818
LANDS GRANTED TO NAWAB OF JAORA-TREATY OF MANDSAUR, 1818.
CEDED TO KOTA-1818
TRIBUTARY STATES OF INDORE
STATE TERRITORIES 1818
0 50 100 150 200
Kilometers

In order to make the Holkar dominions more compact, various exchanges of territory were affected between 1861 and 1868 (Figure 6), the districts of Satwas in Nemawar, and Barwaha, Dharangaon, Kharwad, and Mandleshwar in Nimar, being exchanged for lands held in the Deccan, Uttar pradesh and elsewhere. The actual transfer of lands was finally concluded in 1868; but the adjustment of the accounts in connection with the exchanges was not finally settled till 1881, when a kharita (Appendix IV) was sent to the Maharaja, intimating the conclusion of the negotiations and approval of the arrangements that have been made.[33] From 1864 to 1878, the Maharaja had disputed the interpretation of the sixth article of the Treaty of Mandsaur (1818) by which he claimed certain territories in and south of the Satpura range. A boundary commission demarcated a line in 1864, which was not acceptable to the Maharaja. After a further, full and complete inquiry, some 932 km^2 of territory was made over to him in 1878, and the case was treated as being finally closed.[34]

In 1870-71 an exchange of territory was negotiated between the States of Indore and Jhabua, whereby Indore acquired the town and pargana of Petlawad; and Jhabua, the town and pargana of Thandla.[35]

Tukoji Rao II died on 17th June, 1886 and was succeeded by his eldest son, Shivaji Rao. Maharaha Shivaji Rao abdicted in favour of his son, Tukoji Rao III on 31st January, 1903.

Under Tukoji Rao III the post of separate Resident, attached to the Holkar State, was abolished in 1916 and the State came to be in direct relationship with the Agent to the Governor-General in Central India. He, too like his father, abdicted in favour of his minor son, Prince Yashwant Rao Holkar.

Maharaja Yashwant Rao Holkar II was invested with full ruling powers on the 9th May, 1930. He got the Manpur pargana of the British India transferred to Holkar State. In Maharaja's absence from September 20, 1942 to May 15, 1943, while he went for treatment to the U.S.A., Maharaja Vikramsinh Rao of Dewas Senior managed the affairs of the State.[36]

During the World War II, the Maharaja wrote a letter to the President of the United States stressing the imperative need of satisfying nationalist demand in India. It put him into trouble with the Political Department and he retired into his shell. He tried to evolve a 'third force' out of states in the wake of Indian Independence, as he was opposed to the accession of his State. The Maharaja did not attend the meeting of the

MAP 6

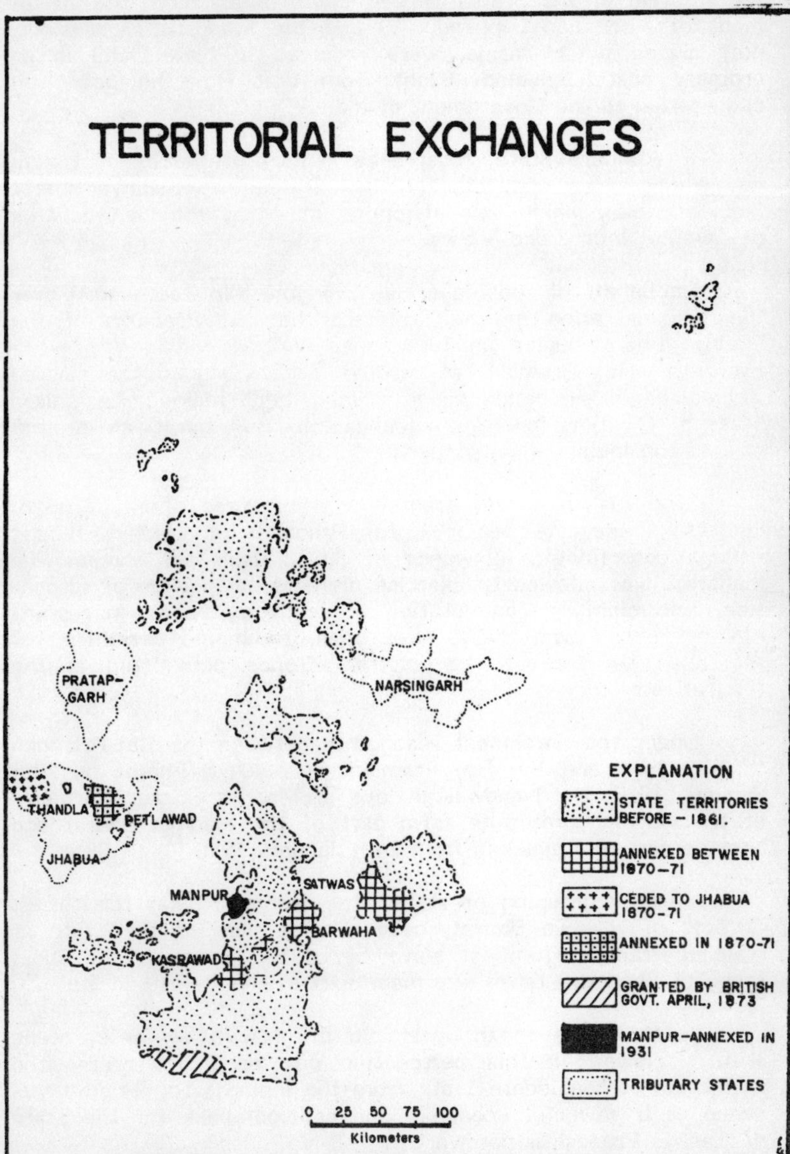
TERRITORIAL EXCHANGES
PRATAP-GARH
NARSINGARH
THANDLA
PETLAWAD
JHABUA
MANPUR
SATWAS
BARWAHA
KASRAWAD
EXPLANATION
STATE TERRITORIES BEFORE - 1861.
ANNEXED BETWEEN 1870-71
CEDED TO JHABUA 1870-71
ANNEXED IN 1870-71
GRANTED BY BRITISH GOVT. APRIL, 1873
MANPUR-ANNEXED IN 1931
TRIBUTARY STATES
0 25 50 75 100
Kilometers

rulers and states' representatives on 25th July, 1948. He had an interview with Lord Mountbatten subsequently, but did not commit himself either way with regard to accession. But to the surprise of the States Department officials, both the Instrument of Accession (Appendix V) and the Standstill Agreement, duly signed by Maharaja, were received in New Delhi in an ordinary postal envelop. Right from that time he gave full cooperation to the Government of India.[37]

The Indore State, thus, became part of the Indian Union, and was coupled with other Malwa States into Madhya Bharat (Appendix I), which was officially known as the United State of Gwalior, Indore and Malwa.

On behalf of the Maharaja his Prime Minister handed over the administration of the State to the Raj Pramukh of the Madhya Bharat Union on 16th June, 1948 (Appendix VII). However, the Raj Pramukh of Madhya Bharat signed the revised Instrument of Accession on 19th July, 1948, making the United State of Gwalior, Indore and Malwa (Madhya Bharat) an integral part of the Indian Union (Appendix VIII).

The Maharaja was granted a privy purse of Rs.15 lakhs. He was made the Senior Upraj Pramukh of Madhya Bharat with a consolidated allowance of Rs.2.5 lakh per annum. The Maharaja was allowed to exercise his personal powers of suspension, remission, or commutation of death sentences, in respect of any person who might have been, or should hereafter be, sentenced to death for a capital offence committed in the territories of Indore.[38]

Under the agreement reached in between the Raj Pramukh of Rajasthan and the Raj Pramukh of Madhya Bharat on 20th January 1950, the Nandwai enclave of Madhya Bharat (formerly of the Holkar) ceased to form part of the Madhya Bharat and became part of Rajasthan from 24th January 1950.[39]

After regrouping of states, the former Holkar territories, as part of Madhya Bharat, have become part of the State of Madhya Pradesh from 1st November 1956. However, the Sunel tappa of Bhanpura tahsil was made part of the Rajasthan State.[40]

Life moves on the principle of growth, we never stand still.[41] Viewed in this perspective one need not regret the extinction of the Indore State from the Indian map. Its re-emergence as a powerful economic and political base for the State of Madhya Pradesh is noteworthy.

References

1. Jadunath Sarkar, in Preface to Raghubir Sinh, Malwa in Transition - A Century of Anarchy, Bombay, 1936.

2. The Imperial Gazetteer of India, Vol. IX, Oxford, 1908, pp. 334-35.

3. 'Subah' is a Persian term for province, which has been termed 'state' in the Indian Constitution.

4. S.M. Edwards and H.L.O. Garrett, Mughal Rule in India, Delhi, 1956, p. 157.

5. Malav Itihas Mandal, The Mandlik Papers and the Family, Indore, 1946, p.11.

6. 'Jai Singh asked Dado Bhimsen, the Peshwa's Agent in the North (in August 1728) to request his master to send the Maratha armies to Malwa at an early date.' Raghubir Sinh, op. cit., p. 161.

7. G.S. Sardesai (Ed.), Selections from Peshwa Daftar, Part XIII, Letter Nos. 17, 23, 25, 27; Raghubir Sinh, op. cit., p. 164; G.S. Sardesai, New History of the Marathas, Vol. II, Bombay, 1948, p. 102.

8. V.V.Thakur, Holkarshahichya Itihasanchi Sadhne, Vol. I, New Delhi, 1970, p. 146.

9. Sardesai (Ed.), op. cit., Part XXX, No.317, p. 300.

10. ibid., No.318, p. 305.

11. Thakur, op. cit., No.26, pp. 16-17.

12. ibid., p. 41.

13. L.C. Dhariwal, The Indore State Gazetteer, Vol. 1, Indore, 1931, p. 22. Sawai Jai Singh die in 1743. He left two sons, of whom the eldest was Isri Singh, but the younger son, Madho Singh, claimed the gaddi of Jaipur on the ground that he was the child o a princess of the House of Udaipur. Malhar Rao helped Madho Singh in gett°ng the gaddi of Jaipur.

14. Malcolm, op. cit., p. 124.

15. Sardesai, op. cit., p. 421.

16. S. Playne, Indian ates - A Biographical, Historical and Administrative Survey, London, 1921, p. 76.

17. Malcolm, op.cit., pp. 147-48.

18. Thakur, op. cit., Vol.II, No.13, p. 163

19. Malcolm, op. cit., p. 194.

20. The Imperial Gasetteer of India Vol. XIII, Oxford, 1908, p. 336.

21. Malcolm, op. cit., p. 195.

23. ibid., p. 202; Jsdunath Sarkar, Fall of the Mughal Empire, Vol. IV, p. 168.

24. ibid., p. 169.

25. The Imperial Gazetteer of India, op. cit., p. 337.

28. Sardesai, op.cit., Vol. IV, p.437.

29. Malcom, op. cit., p. 283.

30. The title of awab was bestowed on Amir Khan by Yashwant Rao Holkar in 1798. ibid., p.206.

31. C.U. Aitchison (Comp.), Treaties, Engagements and Sanads relating to India and Neighbouring Countries, Vol. IV, Calcutta, 19i33, pp. 30-31.

32. Martand Rao was adopted by Malhar Rao's mother, Kesri Bai, and his wife after his death.

33. Aitchison, op. cit., p.17.

34. Dhariwal, op. cit., p. 41.

35. Aitchison, op. cit., p.18.

36. M. Malgonkar,Puars of Dewas Senior, Bombay, 1963, pp. 306-7.

37. V.P. Menon, The Story of the Integration of the Indian States, Calcutta, 1956, pp. 119-20.

38. ibid., pp. 234-38.

39. Government of India, Ministry of States, White Paper on Indian States, Delhi, 1950, Appendix LIII, pp. 362-64.

40. Government of India, The State Reorganisation Act, 1956 (Act 37 of 1956), Section 9.

41. Sardesai, op.cit., Vol. III, p. 519.

CHAPTER 4

Ecological Setting

Whole of the area of the erstwhile Indore State is almost part of the Central Highlands, a compact block of mountains, hills and plateaus, interspersed with valleys and basins.[1] It is an area of great physical complexity and lies almost in the heart of India. Its scarped plateaus and troughs represent the buckling of the northern foreland of the peninsular block under the stress of the Himalayan mountain-building.[2]

In the geomorphological evolution of the area, apart from the general opposition of the Gondwana block and the Himalayan folding, the ancient Aravalli in the north-west, the gneissic block of Bundelkhand in the north, and the Deccan tableland in the south have played very importnat roles in controlling the general lay-out of the plateaus and troughs squeezed between them.

The old Indore State fall under following structural unit[3] from south to north :

(a) the Satpura block ;
(b) part of the Narmada-Son furrow ;
(c) the Vindhyan Hills (Deccan lava) ;
(d) the long slopes of the Deccan larvas in the Chambal basin, north of the Vindhyan Hills ; and
(e) the Chambal-Sindh basin, north-east of Malwa.

PHYSIOGRAPHIC REGIONS

On the basis of the physical components the State can be divided into following physiographic regions (Figure 7) :

1. The Malwa Plateau;
2. The Vindhya Range ;
3. The Narmada Valley ;
4. The Satpura Range ; and
5. The Chambal-Sindh Basin.

MAP 7

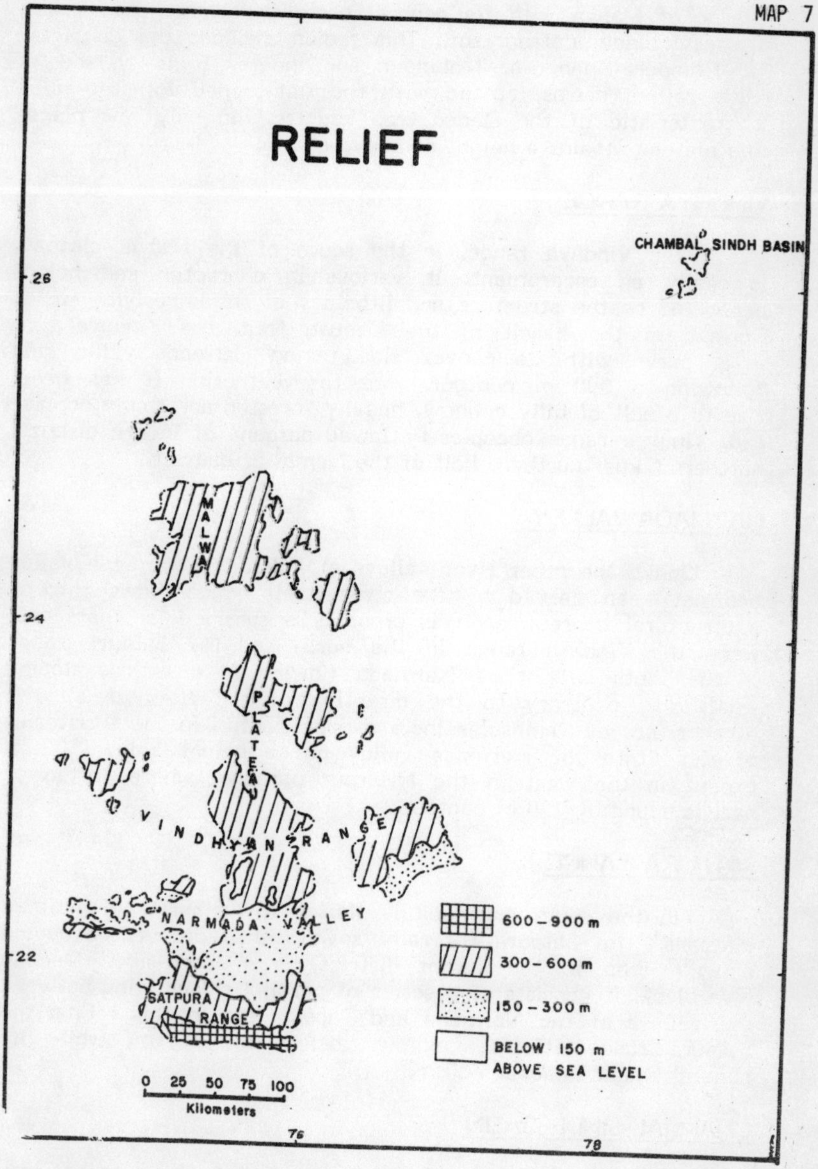
RELIEF
CHAMBAL SINDH BASIN
26
24
22
76
78
MALWA
PLATEAU
VINDHYAN RANGE
NARMADA VALLEY
SATPURA RANGE
600-900 m
300-600 m
150-300 m
BELOW 150 m
ABOVE SEA LEVEL
0 25 50 75 100
Kilometers

MALWA PLATEAU

To the north of the Vindhya range occurs the extensive lava plateu of Malwa with its general northward slope, good drainage[4] and black 'cotton' soil. This region includes the districts of Rampura-Bhanpura, Mahidpur and indore. It is a land of wide rolling downs, dotted with the flat-topped forested hills, characteristic of the Deccn trap country.[5] In only few places the plateau attains a height of above 600 m.

VINDHYA RANGE

The Vindhya range, in the south of the Malwa plateau, is really an escarpment. It various in character and height depending on the structure and lithology of the underlying rocks. Formed of the basalt of the Deccan trap, the Vindhya runs in a curve with its convex side facing Narmada valley and following a 300 m contour. Due to weathering it has given rise to a belt of hilly country, heavily forested and thinly populated. Vindhya range occupies Petlawad pargana of Indore district, northern Nimar northern half of the Nemawar district.

NARMADA VALLEY

Unlike the other river valleys of India, the Narmada valley has not been carved by the river itself which flows through a structural depression. It is probably a subsided landmass between the Vindhya range in the north and the Satpura range in the south with river Narmada flowing in a valley, sloping westwards, contrary to the direction of the flow of all the other rivers of Peninsular India except Tapti.[6] In the territories of this State the Narmada valley has a height below 300 m except in the west in the Nisarpur pargana, where its lower part is bound by 150 m contour.

SATPURA RANGE

In the south Narmada is flanked by a series of scarped plateaus. The Satpura (literally seven folds) proper, occupying a wide area in the Nimar district, is a steep-sided Deccan lava block. It consists of a series of parallel ridges lying between the valleys of the Narmada and Tapti. Here in this region the Satpura range attains maximum height of 900 m, while its general height is about 600 m.

CHAMBAL-SINDH BASIN

In the northern part of the Chambal-sindh basin, in between

Malwa plateau in the south and river Yamuna in the north, lies the Alampur pargana of the Mahidpur district. As part of the Gwalior plain, a virtual extension of the Ganga plain, it has a levelled aluvial topography with an average height of 150 m.

GEOPOLITICS OF THE HILLS

The hill system of the State is formed of the Vindhya range and its several branches, and the Satpura. The main Vindhya range, forming the abrupt termination to the Malwa plateau, runs across the State roughly from east to west with an average elevation of 600 m above the sea level. It enters the State in the Khategaon pargana of the Nemawar district at 77°3'E, forming the northern and north-western boundaries of the district. After a short span of old Dhar's territory the Vindhya rappers on the map in Indore district, and follows a west-south west course upto Maheshwar in Nimar district, and then takes a northward bend. It makes the western boundary of the Petlawad pargana. The Satpura range enters the state territory at 76°5'E and traverse it for a distance of 112 km with an average breadth of 48 km.

Numerous passes (Figure 7) traverse the Vindhyan scarp. Mention can be made of the following passes, locally referred to as ghats- (1) the Dhantalao ghat in Kannod pargana, leading to Indore from nemawar ;[7] (2) Bhaironghat, through which passes the Indore-Khandwa road; (3) the Jamghat or Jam Darwaja, surmounted by the gateway built by Ahilya Bai ; (4) the Garaghat (also called Ghoraghat and Manpur Ghat), providing passage to the famour Bombay-Agra Road, the Naitonal Highway through which the Marathas entered Malwa.

In the Satpura the only pass of importance, in the State's territory, is Gwalanghat, better known as the Sendhwa pass, through which the National Highway No.3 enters from the Tapti valley, and which even in the pre-British days was a principal route between the North India and Deccan.

In historical geography of Malwa these passes have played very improtant roles. As the passageway between the north and south india Malwa constantly changed hands, 'the invariable apanage to the domains of every monarch, native or barbarian, who became the mastr of the Gangetic plain.'[8] And with the change of rulers, every king holding this region used to be very cautious to keep a close watch on these passes. They established forts as check posts on these routes. In 18th and ealy 19th centuries these forts provided shelter expeditions.

MAP 8

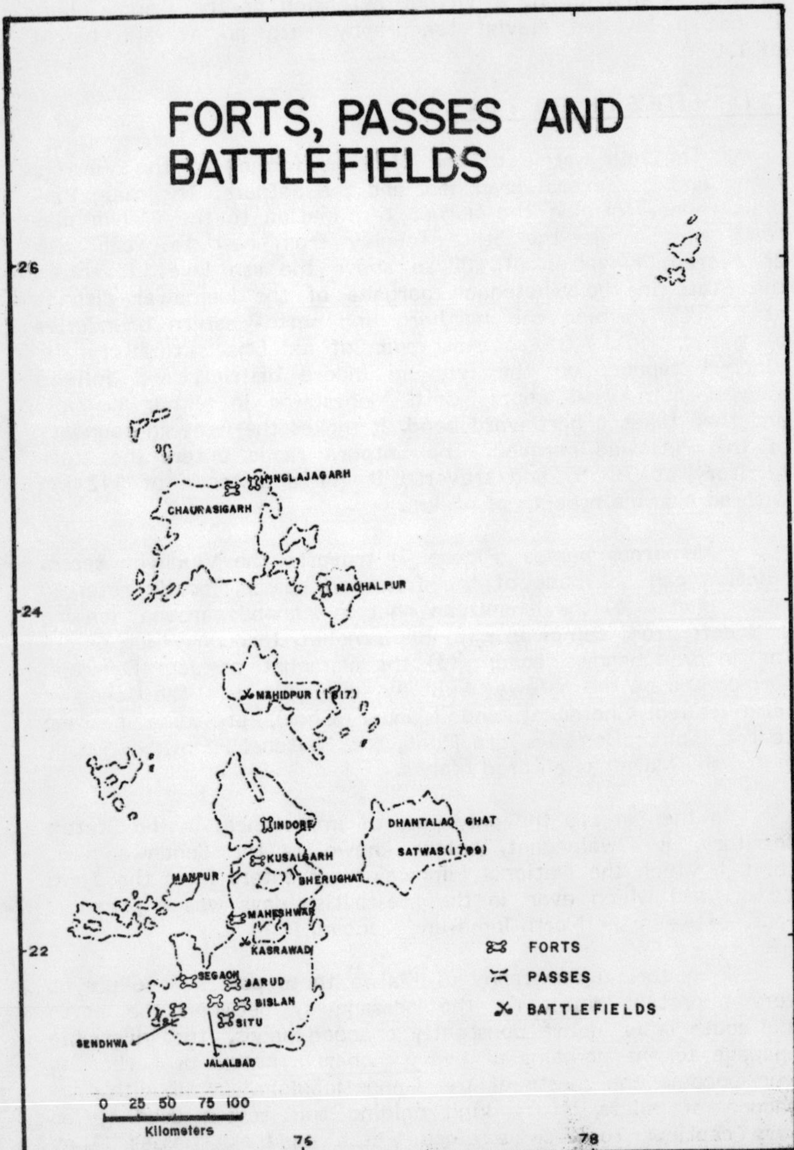
FORTS, PASSES AND BATTLEFIELDS
HINGLAJAGARH
CHAURASIGARH
MACHALPUR
MAHIDPUR (1817)
INDORE
DHANTALAO GHAT
SATWAS (1799)
KUSALGARH
SHERUGHAT
MANPUR
MAHESHWAR
KASRAWAD
SEGAON
JARUD
BISLAN
SITU
SENDHWA
JALALBAD
FORTS
PASSES
BATTLEFIELDS
0 25 50 75 100
Kilometers
26
24
22
76
78

The fort of Kushalgarh (22°28'N-75°50'E) in Mhow sub-division of Indore district is still in a good condition. On the hill to the north-east of Rampura stand the forts of Hinglajgarh, Indergarh and Chaurasigarh, once formidably strong, now in ruins.

Various forts (Figure 7) lie in the Satpura also. Bijagarh and Sindhwa have witnessed the glorious past when Bijagarh was the chief town of a sarkar[9] in Akbar's days, but now they have been reduced to the importance of historical ruins.

DRAINAGE PATTERN

The plateau section of erstwhile Indore State is dissected by a number of rivers and their tributaries (Figure 9). Principal routes - trade and military, between the North India and Deccan follow the valley of main rivers. Most of the rivers (except Narmada) flow from the northern slopes of the Vindhya range. The Chambal and her tributaries drain into the Yamuna in the Gangetic plain. River Mahi flows to the north-west through Rajasthan and Gujarat into the Gulf of Cambay. The tributaries of the Narmada in the Nemawar district (rivers Chandkesar, Dhatuni, Gomi, Bagdi, and Jamner) flow from north across Vindhya range to the south into it. In the north of the Narmada the river system exhibits a radial pattern, in which from the Vindhyan scarp rivers radiate in three directions - north, north--west and south. Rising from the Satpura range, the rivers Bankud, Kunda, Hatni, Borad and Deb drain into the Narmada in the north.

The principal rivers of the State fall under the following four systems, of which the first two are important :

1. The Chambal system which includes besides the Chambal, her affluents the Kalisindh, Chhapi and Ahu; the Chhoti Kali Sindh, Sipra, Gambhir and Khan ;

2. The Narmada system includes the tributaries of Narmada, viz., the Hatni, Borad, Kunda, Bankud, Deb, Chandkesar, Dhatuni, Bagdi, Jamner and Gomi;

3. The Mahi ; and

4. River Sonai of the Sindh-Pahuj system drains the Alampur pargana in the Chambal Sindh basin.

Among the rivers of the state only Narmada is navigable throughout the year between Mortakka and Maheshwar, and for most part of the year from below the Sahasradhara falls

MAP 9

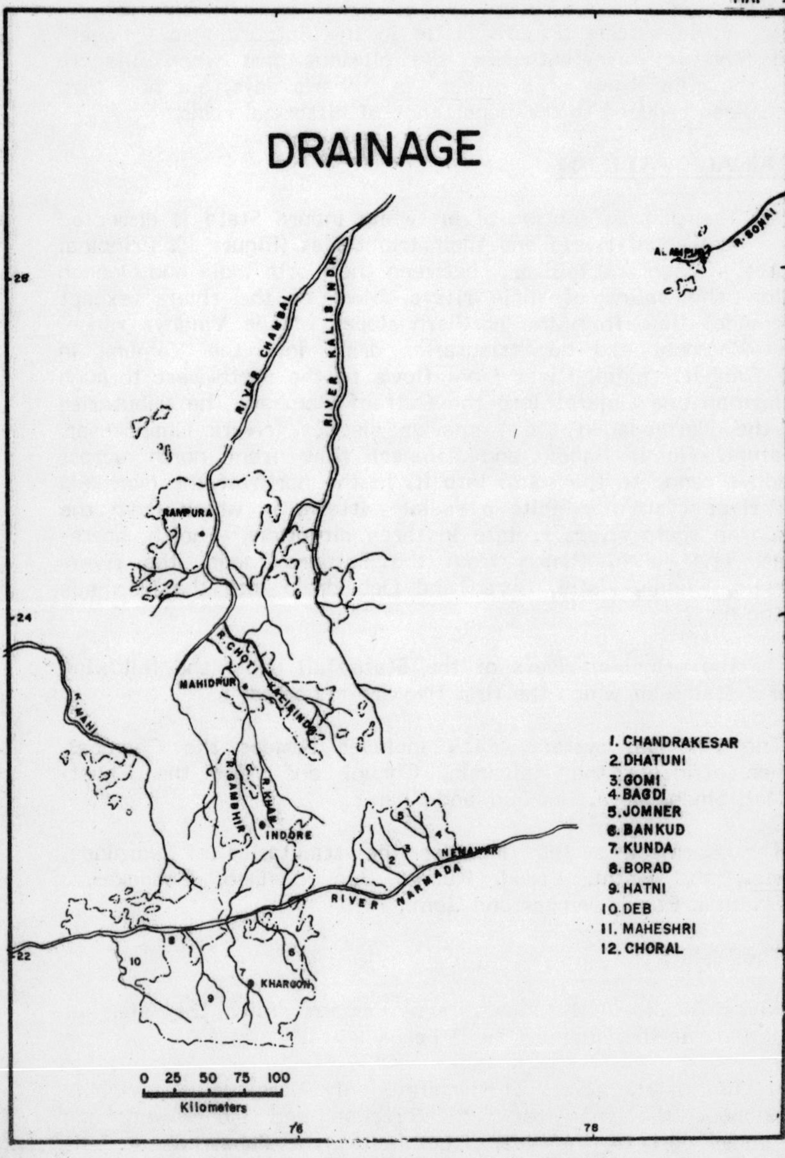
DRAINAGE
RIVER CHAMBAL
RIVER KALISINDH
R. SONAI
ALAMPUR
RAMPURA
R. CHOTI KALISINDH
MAHIDPUR
R. MAHI
R. GAMBHIR
R. KHAN
INDORE
NEMAWAR
RIVER NARMADA
KHARGON
1. CHANDRAKESAR
2. DHATUNI
3. GOMI
4. BAGDI
5. JOMNER
6. BANKUD
7. KUNDA
8. BORAD
9. HATNI
10. DEB
11. MAHESHRI
12. CHORAL
0 25 50 75 100
Kilometers
26
24
22
76
78

to Haranphal. None of the larger rivers is of much use for irrigation owing to the excessive steepness of their banks, though some of their tributaries are used for irrigation.

There are some beautiful waterfalls in this region. The Mendikund and the Patalpani falls in Choral, near Patalpani railway station on Ajmer-Khandwa line of the Western Railway, the Sahasradhara falls in the Narmada near Maheshwar, the Gidh Khoh near Khandel in Indore pargana, the Kundi and Nil-jhari falls ner Sirwel in Khargone pargana are the most important falls in the territories formerly held by the Holkars.[10]

The Chambal, having a fall of over 650 m, flows over rocky surface in the territory of this State. Its banks are 0.3 to 28 m high. With a general width of about 760 m, the Chambal passes through a narrow gorge of 180 m at Chaurasigarh, where it enters the territory of the old Indore State. The first dam of the multi-purpose Chambal Project has been constructed near Chaurasigarh and is named after Mahatma Gandhi.[11] The Gandhi Sagar Dam is 533 m long and 61 m high above the river bed. It holds about 6.85 milion acre-feet (9.8 lakh hectare metre) ofwater. The surface area of the lake measures about 290 km^2. The Gandhi Sagar Dam and power station, completed in 1960, provides irrigation facilities to otherwise unirrigated tracts and deliverspower in bulk to Indore and neighbouring towns and cities in the Central India. This project is a fine example of post-Independence planning in India. Wth two other dams and power stations of Rana Pratap Sagar and Kota Barrage, the Chambal Project hasd got an ultimate capacity of 2 lakh kw of power, sufficient to revolutionize the economic life of this region.

The Narmada Valley Project will provide enormous incentive for the economic development of the Nimar district and the neighbouring tracts.

BOUNDARIES OF THE STATE

The rivers and the relief features had marked the state boundaries to much extent (Figure 3). River Narmada was an effective boundary between the territories of the Indore State on one hand and the States of Barwani and Dhar; and the then British Indian districts of Nimar (Khandwa) and Hoshangabad on the other. River Sipra separate Indore and Dewas. Sunel sub-division was bound by river Kalisindh in the east and river Ahu in the west, thus separating it from erstwhile Jhalawar, Tonk and Kota States. River Kalisindh also marked the boundary between the Jirapur pargana and Shajapur district of the Gwalior state. River Mahi between petlawad pargana and Sailana State; and river Chambal between Garoth sub-division of Rampura--Bhanpura district and jaora State and Mandsaur district of Gwaliur State acted as effective physical boundaries.

The Vindya range separated the Indore State from Bhopal in Nemawar district, Dhar in Indore district as well as the Nisapur pargana of Nimar district (Figure 3).

Most of the state boundaries, except the river and mountain boundaries, had been of arbitrary nature and were superimposed. These boundaries separated the conquered territories between the Maratha generals, who later on became independent of the Peshwa and administered their respective states. A number of enclaves and exclaves had provided the Indore State a complex boundary line, not very easy to measure.

The state boundaries, after the Treaty of Mandsaur in 1818, did not cause any serious friction with the neighbouring states as the Holkars had always cordial relations with their neighbours and also with the British Government.

CLIMATE

The climate in its direct effects on human beings has often been stressed, perhaps overstressed.[12] It determines the kind of food man can raise, and it strongly affects hius mode of life. The cultural attainments of man and his vitality are influenced by the climate. The development of political power, too, is closely related with the climate.

Cold, heat and aridity are the three climatic factors that have a limiting effect on human occupancy and the development of political power.[13] These factors more or less remain the same in whole of the erstwhile Indore State territories. Abrupt fluctuations or changes in the climatic conditions have been rarity in the region.

TEMPERATURE

The Tropic of Cancer divides the territory of the State into two equal parts-warm temperate and tropical. But on account of the monsoonal character of the Indian climate as a whole, and upland nature of the Central Highlands, very little effect is produced by the Tropic of Cancer in the distribution of temperature in the State.

Temperatures decrease, in general, with the increasing altitude. During the cold weather the plateau districts of Rampura-Bhanpura, Mahidpur and Indore record an average temperature of about 12.8°C, while the Narmada valley districts of Nimar and Nemawar register 14.3°C each. The summer season is extremely hot with average maximum temperatures[14] of 33.9°C in the plateau region nd 35.0°C in the valley region. Table 4.1 shows the range of minimum and maximum temperature in various districts of the State.

Table 4.1 : Minimum and Maximum Temperature (°C)

District		Elevation above sea level (m)	Cold weather average minimum (°C)	Hot Weather average maximum (°C)
1.	Indore	556	12.9	33.3
2.	Mahidpur	518	12.8	34.3
3.	Nemawar	320	14.3	35.1
4.	Nimar	320	14.3	35.1
5.	Rampura-Bhanpura	497	12.8	34.3

Source: L L.C. Dhariwal, The Indore State Gazetteer, Indore, 1931, Vol. 1, p. 18.

The region has a marked winter season as the January temperatures average below 17°C and June temperatures above 30°C representing hot weather. The average variation in temperature from the warmest to the coldest month is about 14°C. The mean monthly tempeatures for Indore City represent very well the general temperature conditions in the State (Table 4.2).

RAINFALL

The south-west monsoon, which suddenly breaks about the first half of June and lasts for about four months, accounts for about 90 per cent of the annual rainfall. Table 4.2 shows very clearly the paucity of rainfall during the winter months and its abundance during the rainy season.

The distribution of rainfall is alike in almost all the territories of the former Holkar State, which records an annual average of 73.38 cm. The districts of Nimawar and Nemawar in the Narmada valley get comparatively more rainfall than the plateau districts of Indore, Mahidpur and Rampura-Bhanpura (Table 4.3 and Figure 10.)

Table 4.2 : Mean Monthly Temperature, Rainfall, Humidity, Wind Velocity and Direction at Indore City

Climatic phenomena	Jan	Feb	Mar	Apr	May	Jun	Jul	Aug	Sep	Oct	Nov	Dec	Total/ Mean
Temperature(°C)	17.6	19.4	23.8	26.9	31.5	30.9	27.6	25.9	26.0	23.3	20.1	17.6	24.2
Rainfall(cm)	0.71	0.46	0.15	0.36	1.47	14.48	25.07	20.35	16.59	2.92	0.94	0.53	84.03
Relative Humidity (%)	60	49	40	39	50	72	85	88	84	61	56	59	62
Wind velocity (Km.p.h.)	3.0	3.7	4.3	5.6	8.2	9.0	8.4	7.4	5.2	2.6	2.2	2.2	5.2
Wind Direction	N36E	N29W	N37W	N45W	N71W	S87W	S85W	S85W	N60W	N18E	N45E	N68E	N66W

Source : L.C. Dhariwal, The Indore State Gazeteer, Indore, 1931, Vol. - III, P. 66-67.

MAP 10

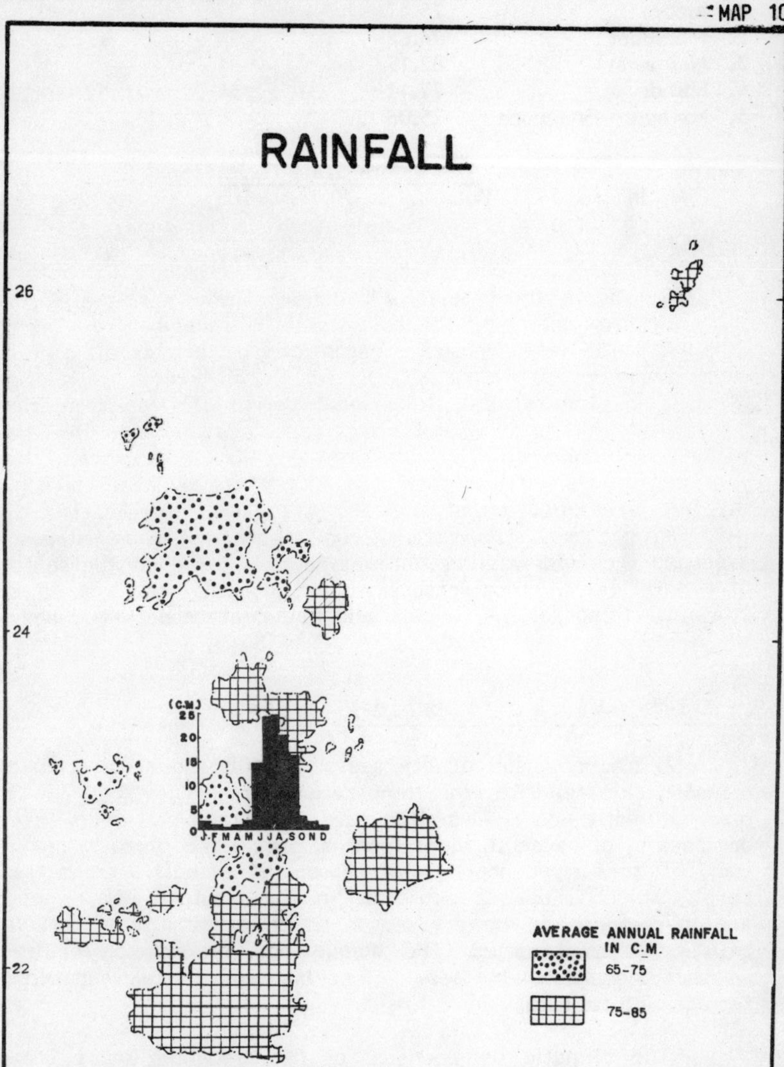
RAINFALL
26
24
22
76
78
(C.M.)
25
20
15
10
5
0
J F M A M J J A S O N D
AVERAGE ANNUAL RAINFALL IN C.M.
65-75
75-85
0 25 50 75 100
Kilometers

Table 4.3 : Average Annual Rainfall by Districts

District	Rainfall (cm)	Years of observation
1. Indore	75.67	37
2. Mahidpur	76.68	34
3. Nemawar	82.19	34
4. Nimar	77.11	34
5. Rampura-Bhanpura	75.06	34

Source : L.C. dhariwal, The Indore State Gazetteer, Indore, 1931, Vol. III. pp. 70.71.

SEASONS

Climatically the year falls into four seasons : (a) the North--East Monsoon season or the cold weather (December to March); (b) the South-West monsoon season or the season of general rains (June to September); (c) the transitional hot weather (April-May); and (d) the transitional period of the retreating South-West Monsoon (October-November), recognized by the Indian Metropological Department. Popularly, however, the year can be divided into the 'cold wether' (November-February), the hot weather (March-May), and the rains (June-September), with October as a sticky transitional month, the rains not quite over and the cold weather not yet set in .[15] Except the season of general rains, other seasons remain generally dry with lower or higher temperatures in cold and hot weather periods respectively.

CLIMATE AND POLITICO-GEOGRAPHIC DEVELOPMENT

A longer period of dry season in this region of tropical monsoon climate, with high temperature for most of the months, gives climatic conditions that are favourable to man. This dry season, of seven to eight months, breaks the climatic monotony of the pronounced rainy season. It affects the vegetal cover, which is not as dense as in the tropical wet regions, and in response to the dry season which exhibits some drought resisting characteristics. The soils, too, respond to the dry weather conditions by lesser leaching and better suitability for agricultural uses.

The climatic homogeneity of the erstwhile Indore state was an asset to the political unity of her territories. Everywhere in the state, due to uniformit of her climate, people spoke the same 'geographical language.'[16] Becase of the similar environmental conditions they could understood one anothers' point of view more easily and comfortably. Accordingly it was easier to administer the isolated tracts of the state from one central point. Thus, the homogeneity of climate was responsible, in part, for the fruitful existence of the Indore state.

The climatic affinities between this region and Maharashtra have played an important role in the evolution of the Holkar state in the Central Highlands of India alongwith other Maratha states like Gwalior, Dhar and Dewas. It was the climate of Malwa region, which attracted the Marathas not only for invasion of this old Mughal province but also to settle down as rulers.

SOILS

Soil, like water and air, is an integral part of the natural landscape, and it affects the economy of a state. All our food, even sea-food indirectly, comes from the soil either in the form of plants that we eat, or animals which provide us with meat.[17] Besides food, we get minerals and fibres from soil.

The state's power potential has a firm base in her soils. 'When the soil is lost, prosperity and culture of the country are also lost.'[18] Therefore the conservation of the soil to maintain its fertility, in order to keep economy sound, is one of the important duties of the states.

SOIL TYPES

Most of the state territories lie in the wide belt of fertile regur soils (Figure 11), which have been derived from the old lava deposits. As most of the region is important cotton growing tract, this soil is sometimes referred to as 'Black Cotton Soil.' With a depth of 6 to 15 metres, the regur soils are very fertile due to moisture-retentive qualities. They are rich in iron, lime and alumina but lack in phosphorus and organic matter. These soils bear annually two first rate crops of all the ordinary grains without irrigation.

Red soils cover most of the hilly areas of the state in the vicinity of the Vindhya and Satpura ranges. They are poor in nitrogen, phosphorous and humus as well as in lime. Their low fertility is responsible for the growth of autumn crops only in the hilly tracts of the state.

The alluvial soils, with adequate lime, potash and phosphoric acid, are very fertile in the older flood plain[19] of the trans--Yamuna region, where Alampur sub-division is situated. The

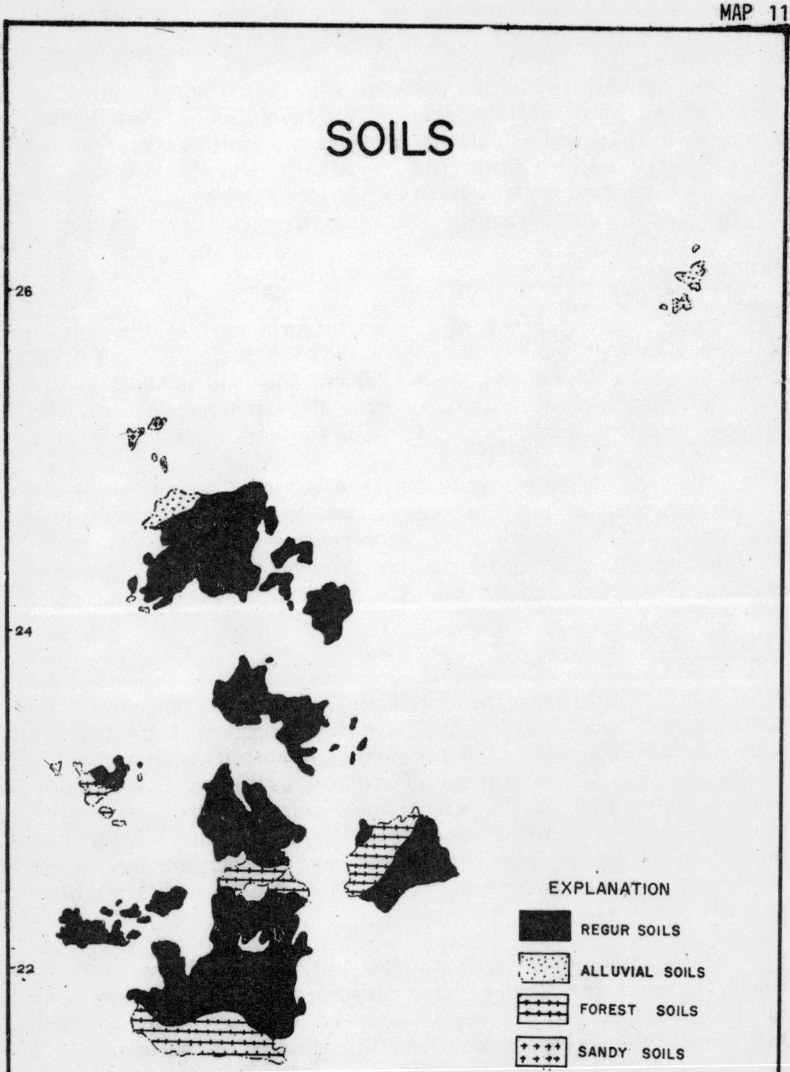
MAP 11
SOILS
26
24
22
76
78
EXPLANATION
REGUR SOILS
ALLUVIAL SOILS
FOREST SOILS
SANDY SOILS
0 25 50 75 100
Kilometers

alluviums also cover narrow strips along the river valleys in the plateau section. Locally known as talakh in Nimar, the fertile alluvial deposits bear excellent crops of tobacco and vegetables as well as of other grains.

The soils of the State are generally of the zonal order as they include 'those mature soil types which have well developed characteristics, based largely upon the dominance of climate and vegetation among the soil forming processes.[20] Latosols or red soils, covering the hilly tracts, are pedalfers (ped means earth or ground, al indicates aluminum, and fer for iron), while the chernozemic black soils fall in the group of pedocals (ped means earth, and cal stands for lime or calcium), which provide the best agricultural lands to the people. The azonal[21] alluvial soils cover very small area in the State, but are of prime agricultural importance.

SOIL CONSERVATION

'When the soil is gone, men must go and the process does not take long,' emphatically remarked President Theodore Roosevelt in his 8th annual message to the American Congress. The soil erosion has been described as 'a disease of civilization but one so serious that it threatens the very extinction of mankind.'[22] Soils, which take thousands of years for development, once eroded cannot be replaced. Therefore, there is a need of preservation and conservation of the existing soils.

Soil conservation in its widest sense includes not only control over soil erosion but all those measures like correction of soil defects, application of manures and fertilisers, proper crop rotations, irrigation, drainage etc. which aim at maintaining the productivity of the soil at a high level. In the low-lying Alampur pargana destructive soil erosion has been checked by constructing dams.

The fertile soils of the state had been an asset to the Holkars, who never visualized any serious famine within their borders. Indore remained a food-surplus state right from its inception. Abundance of food, grown in the Holkar territories, was the most important economic factor helpful in her prosperity. Her leading position among the then princely states was a direct result of her soils-their fertility and productivity.

NATURAL VEGETATION

Forests, the most conspicuous form of the natural vegetation, represent 'associations of plants, predominantly trees.

They constitute one of the man's greatest, as well as one of his most sadly misused, resources.[23] Their pattern and production directly affect the economic viability and political stability of state. Dense forests provide shelter to the unsocial elements and political adversaries. But they also provide rich timber and other products needed for economic transformation of the states.

In the erstwhile Indore state the forests covered an area of 6,669 km^2 or 26 per cent of the total geographical area. The state forests had been divided into five divisions for the purpose of administration. The Indore division consisted of the Indore civil district, and the Barwaha division covered the Vindhyan forests of the Nimar district. The Khargone division included the forests on and along the Satpura in the southern half of the Nimar district. Nemawar, while the forests lying in the Rampura-Bhanpura and Mahidpur districts were grouped into the Rampura division(Table 4.4).

FOREST TYPES

According to Champion's classification[24] of Indian forests, thre are two major categories of tropical forests found in the territories of the former Holkar State. These are ; (a) dry deciduous, and (b) thorn forests. Scrubs are found in certain parts of the State (Figure 12).

The zone of tropical deciduous forests is found in most parts of the State except in Rampura division. It includes the

Table 4.4 : Forest Area in the Indore State

	Division	Block	Area(km^2)	Total area (km^2)
1.	Indore	Main Vindhya	544	756
		Other minor	212	
2.	Barwaha	Main Vindhya	578	940
		Other minor	362	
3.	Khargone	Satpura main	2,466	
		Other minor	262	2,728
4.	Memawar	Main Vindhya	757	
		Other minor	205	962
5.	Rampura	Pathar hills	1,099	
		Other minor	184	1,238
	Total		6,669	6,669

Source : L.C. Dhariwal, The Indore State Gazetteer, Indore, 1931, Vol.I, p. 194.

MAP 12

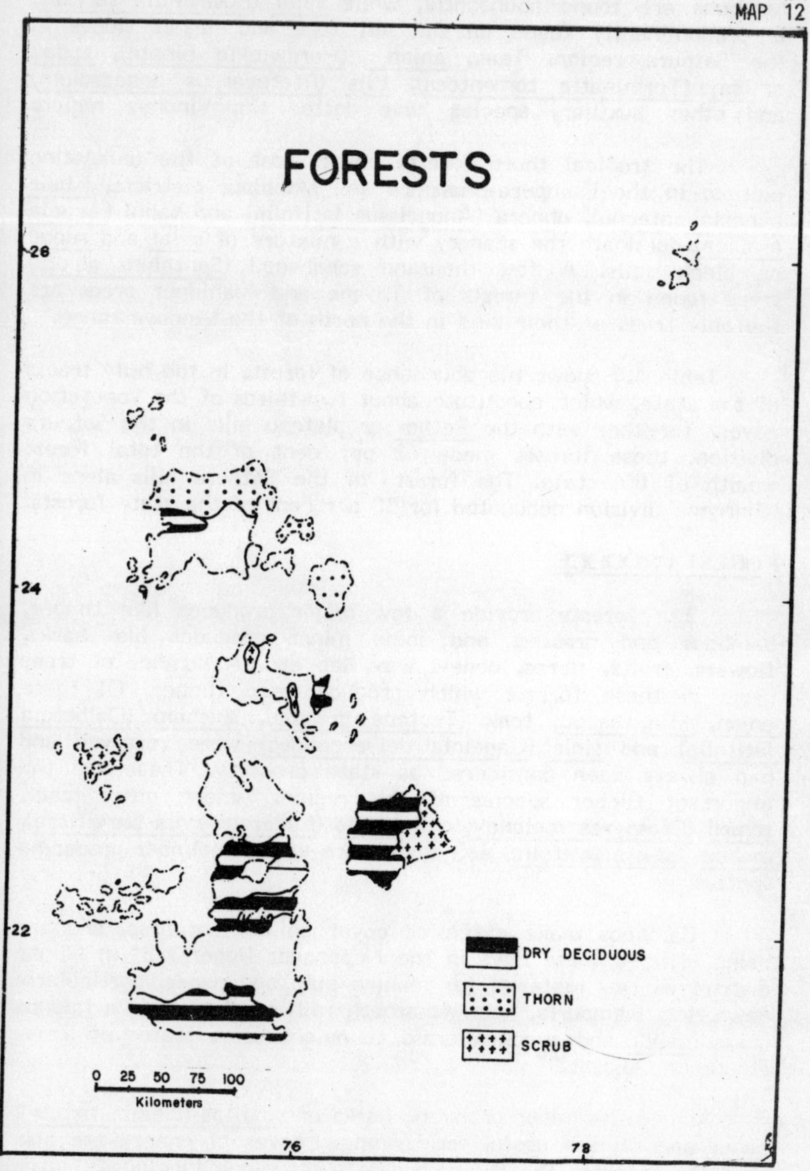
FORESTS
26
24
22
DRY DECIDUOUS
THORN
SCRUB
0 25 50 75 100
Kilometers
76
78

forests of the Vindhya and Satpura ranges and the adjoining areas. On the low-lying hill slopes and the valleys teak and bamboos are found abundantly, while salai (Bosewellia serrata) is predominantly found on the hill tops and higher slopes in the Satpura region. Teak, anjan (Hardwickia binata), sadad or saj (Terminatia tomentosa), bija (Pterocarpus anacardium) and other auxiliary species have dotted the Vindhya region.

The tropical thorn forests cover most of the undulating plateau in the Rampura-Bhanpura and Mahidpur districts. Khair (Acacia catechu), dhaora (Anogeissus latifolia) and babul (Acacia arabica) dominate the scenery with a mixture of salai and sadad on black soils. A few thousand sandlwood (Santalum album) trees found in the forests of Tarana and Mahidpur areas are the only trees of their kind in the north of the Vindhya range.[25]

Table 4.4 shows the abundance of forests in the hilly tracts of the state, which constitute about two-thirds of the vegetation cover. Together with the Pathar or plateau hills in the Satpura division, these forests made 82 per cent of the total forest wealth of the state. The forests of the Satpura hills alone in Khargone division accounted for 38 per cent of the state forests.

FOREST PRODUCE

The forests provide a few major produces like timber, bamboos and grasses; and some minor produces like barks, flowers, fruits, fibres, honey, wax, lac etc. A number of trees grow in these forests which produce good timber. Of these anjan, bija, sadad, teak (Tectona grandis), shisham (Dalbergia latifolia) and tinis (Ongeinia dalbergioides) were reserved and had always been considered as state property. These are the important timber species of this region. Among other trees, temru (Diospyros melanoxylon), lendia (Lagerstroemia parviflora), mahua (Bassia latifolia) and dhaora are valuable timber producing species.

Bamboos make extensive cover on the Satpuras, and are being utilized these days in the Nepanagar Paper Mill in Nimar district as raw material for making pulp and paper, particularly newsprint. Rampuria, near Amargarh railway station, is a famous grass mandi and supplies grass to neighbouring states of Rajasthan and Gujarat.[26]

Among the minor products, barks of amaltas (Cassia fistula), babul and saj are useful for tanning. Leaves of dhaora are also used for tanning. The temru and asta (bauhinea facemosa) leaves are used for rolling up of country cigarattes or bidis; and those of khajur (Phoenix sylvestris) for making brooms, mats, fans, brushes etc.

The mahua fruits yield a pure and valuable oil, which was an important item of export to Europe those days. Baheda (Terminalia bellerica) is used for tanning and for medicine. Babul, bija, baheda and dhaora yield valuable gums, an item of export. Dyes are made from the roots of al (Morinda tinctoria and flowers of dhak (Butea frondosa), bark of acacia and the pods of babul. From the young branches of anjan strong ropes are made. Honey, wax and lac are also collected in small quantities. Besides these major and minor produces the forests also provide the domestic fuel and charcoal. Thus forests are fundamental in the economy of the region.

Psysiography, drainage lines, climate, soils and vegetation have played, directly or indirectly, an important role in shaping the political geography of Central India. These ecological elements influenced, to much extent, the internl structure and external relationships of the erstwhile Indore State in the pre--Independencee days.

References

1. The Gazetteer of India, Vol, 1, Delhi, 1965, p. 40.
2. O.H.K.Spate and A.T.A.Learmonth, India and Pakistan - A General .and Regional Geography, London, 1967, p. 611.
3. ibid., pp. 613-14.
4. 'Districts' here refer to the districts of old Indore State. They have no relation with the present revenue districts of Madhya Pradesh.
5. L.C. Dhariwal, The Indore State Gazetteer, Vol. 1, Indore, 1931, p. 2.
6. The Gazetteer of India. Vol. I, op.cit., p. 44.
7. The region lying between 20°27' and 22°32'N and 75°55' and 75°57'E is made up of three successive passes, viz., (a) the Semrolghat on the Indore-Khandwa Road, on 14th milestone (22nd km point), (b) the Bhaironghat, on 17th milestone (27th km point) and mentioned on the topographical maps, and (c) the Ravighat on 20th milestone (32nd km point). Dhariwal, op. cit., p. 4.
8. Census of India 1931, Vol. XX, Part I-Report, p. 3.
9. Refers to a province.
10. Dhariwal, op. cit., p. 12.
11. R.N. Dubey, Economic Geography of India, Allahabad, 1968, p. 367.
12. Pul Fordham, The Geography of African Affairs, Middlesex, 1965, p. 23.

13. S.Van Valkenburg and C.L. Stotz, Elements of Political Geography, New Delhi, 1963, p. 63.
14. Valley region in this Chapter stands for the Narmada valley.
15. O.H.K.Spate, India and Pakistan-A General and Regional Geography, London, 1960, p. 41.
16. Valkenburg and Stotz, op.cit., p. 71.
17. V.C. Finch, G.T. Trewartha, A.H. Robinson and E.H. Hammond , Elements of Geography - Physical and Cultural, New York, 1957, p. 435.
18. Dubey, op. cit., p. 101.
19. Old flood plains are called bangar in the Ganga valley.
20. Finch, Trewartha, Robinson and Hmmond, op. cit., p. 445.21. Soils not having well developed soil profiles fall under azonal order.
22. L.D. Stamp, Africa - A Study in Tribal Development, London, London, 1953, p. 101.
23. C.F. Jones and G.G. Darkenwald, Economic Geography, New York, 1966, p. 61.
24. Based on Champion's classification as described by Spate, op. cit., pp. 66-69.
25. Dhariwal, op.cit., p. 197.
26. ibid., p. 200.

11 Introduction

9. G. Mac Munn, Indian States and Princes, London, 1936. p.69.
10. 'States' in the present politico-administrative set-up of India signify 'provinces'. The Constitution of India, Delhi, 1950, Article 1.
11. 'Mahal' signifies a sub-division, tahsil or pargana.
12. L.C. Dhariwal, The Indore State Gazetteer, Vol. I, Indore, 1931, p. 412.
13. Surendranath Dube, Census of Central India, Vol. I- Holkar State - Tables, Indore, 1941, p.1.
14. Dhariwal, op. cit., p.339.
15. W.A.D. Jackson, "Whither Political Geography", Annals of the Association of American Geographers, Vol. 48, 1958, p. 178.
16. Derwent Whittlesey, The Earth and the State, New York, 1944, p. 585.
17. Lewis M. Alexander, World Political Patterns, London, 1957, p. 41.

CHAPTER 5

Economic Resources

Economic strength has always been an instrument of political power.[1] Economic and political activities, therefore, go hand in hand in any political area and are very intricately intertwined. A high level of economic production is to a large extent dependent upon the efficiency of the State in maintaining peace and order, in protecting property, in reorganizing initiative, and in rewarding effort.[2]

The major economic resource determining the power potential of a state are : (1) Agriculture and associated activities, (2) Minerals, (3) Industries, and (4) Trade.

AGRICULTURE

Indore was essentially an agricultural state as most of her people were dependent on agriculture for their livelihood. Agriculture provides practically all the food grains consumed within the area and exported to other places. It also yields raw materials like cotton and oilseeds for the principal industries of the Malwa plateau. With all the inherent fertility of her rich blackcotton soil, and the good investment of labour, the yield per hectare is very low. The prevailing subsistence farming with too many of labour units and too little of capital, and still too little of production ensures but a precarious living.

LAND UTILIZATION

The general pattern of land utilization in different districts of the erstwhile Indore State is shown in Table 5.1 (Figure 13). The cultivated area, which averaged only 50 per cent for the State as a whole, differed from district to district, ranging between 35 and 61 per cent. The State had a large proportion of uncultivated land[3] (18 per cent), while the area not available for cultivation, too, was quite large (15 per cent) due to infertile hill tracts and <u>kharab</u>' (bad) lands, not suitable for agricultural purposes.

Table 5.1 : Land Utilization in Indore State

(Figs. in brackets represent percentage) (in thousand ha)

District	Net cropped area	Fallow land	Uncultivated land	Not available for cultivation	Forests	Total
1. Indore	194 (61)	13 (4)	68 (22)	25 (8)	17 (5)	317 (100)
2. Mahidpur	168 (55)	19 (7)	66 (21)	46 (14)	8 (3)	307 (100)
3. Ramapura-Bhanpura	156 (42)	17 (4)	80 (22)	97 (26)	25 (6)	375 (100)
4. Nimar	404 (52)	61 (8)	99 (13)	113 (15)	94 (12)	771 (100)
5. Nemawar	82 (35)	26 (10)	52 (22)	15 (7)	62 (26)	237 (100)
STATE	1,004 (50)	136 (7)	365 (18)	296 (15)	206 (10)	2,007 (100)

Source : Holkar State Administration Reort, 1945, Indore, Appendix II.

Indore district had the highest proportion of the area under cultivation due to its favourable natural factors like moderate rainfall, gentle slope, with fertile blackcotton soil, combined with high levels of cultural and other developments attained by her people. In the Nemawar district only about a third of of its area was under plough, with fairly high percentage of forest lands. Rampura-Bhanpura district, due to her slopes and soil conditions and dry climate, had much more area under the category - not available for cultivation.

Considerable possibilities of extension of cultivation Nemawar and Mahidpur districts exists, and with the multiplicity of demands of the growing population, the area under cultivation will, undoubtedly, increase.

BALANCE OF FOOD

Food is probably the most important of the economic elements and has become a powerful political weapon in a world torn by conflicting philosophies and doctrines. Lack of adequate supply, resulting in undernourishment and finally in starvation[4] has been one of the chief causes of unrest throughout history.

Cereals and pulses constitute the chief food of the people in the State. Jowar and bajra (millets), wheat, maize and gram

MAP 13

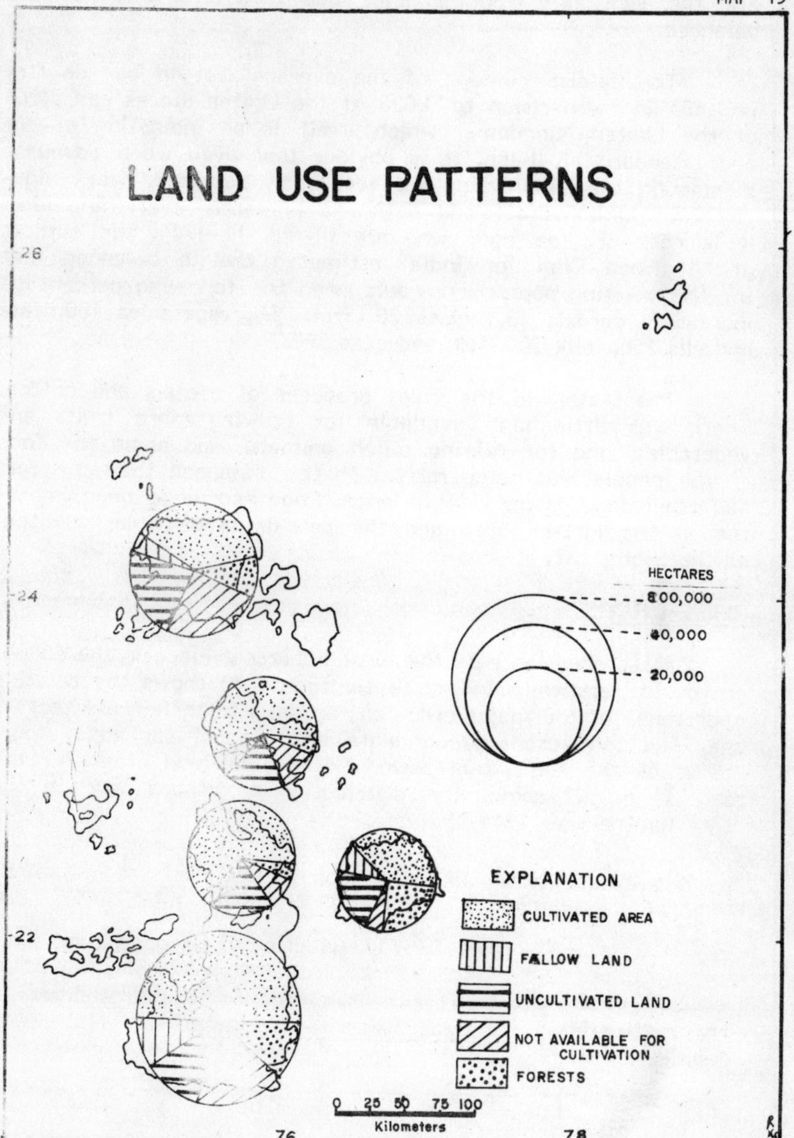
LAND USE PATTERNS
26
24
22
HECTARES
800,000
40,000
20,000
EXPLANATION
CULTIVATED AREA
FALLOW LAND
UNCULTIVATED LAND
NOT AVAILABLE FOR CULTIVATION
FORESTS
0 25 50 75 100
Kilometers
76
78

are the cereals, and tuar, urad, mung and masur are the pulses which are part of daily diet of the people. Vegetables and fruits are the subsidiary food articles but their proportion is not balanced.

The caloric content of the average diet in our country is 1,683 in comparision to 3,088 of the United States and 3,068 of the United Kingdom,[5] which itself is an indication of our poor standard of living. It is obvious that even when adequate in amount, the Indian diets are very badly balanced, over-weighted with carbohydrate and deficient in nearly everything else: it is not only the poor who are ill-fed. In 1944, the authors of 'A' Food Plan for India' estimated that a balanced diet for the existing population would need the following percentage increase : cereals 10, pulses 20, fruits 50, vegetables 100, fats and oils 250, milk 300, fish and eggs 500.[6]

The state was the chief producer of cereals and cotton. There was little land available for growing more fruits and vegetables, and for rearing milch animals. And again the food of the people was comparatively better balanced than the rest and other food grains in abundance. Food had never been imported in the Holkar State,and the cereals used to be an item on the export list.

CHIEF CROPS

Crop production held the most important place in the economy of the erstwhile Indore State. Figure 13 shows the relative importance of the major crops on the basis of their gross hectareage. The food crops predominate and account for about four-fifths of the total area sown and the non-food crops for the rest. Table 5.2 shows the distribution of croppd land in the State for the year 1944-45.

The Indore State was predominantly a Kharif crops dominated area, growing more jowar, bajra, cotton, maize etc. than

Table 5.2 : Distribution of Cropped Land

(In thousand ha)

Gross cultivated area	Rabi	Khariff	Double cropped area	Net Cultivated area
2,113	580	1,533	106	2,007

Source : Holkar State Administration Report, 1945, Indore, Appendix II.

MAP 13A

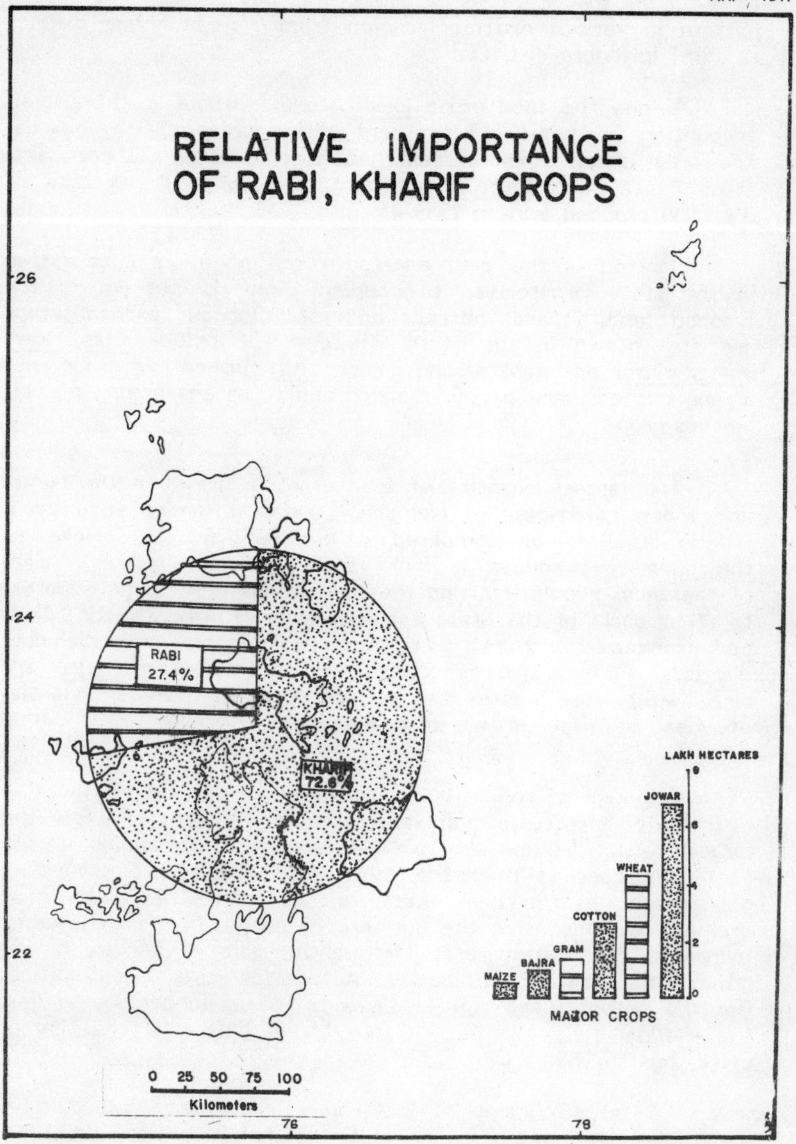

RELATIVE IMPORTANCE OF RABI, KHARIF CROPS
26
24
22
RABI
27.4%
KHARIF
72.6%
LAKH HECTARES
JOWAR
WHEAT
COTTON
GRAM
BAJRA
MAIZE
MAJOR CROPS
0 25 50 75 100
Kilometers
76
78

Rabi crops like wheat, gram etc. The proportion of Kharif to Rabi in the state as a whole was 73:27, with an uneven pattern in various districts ranging from 95:5 in Nimar district to 39:61 in Indore district.

Among the food crops jowar stands without a close rival, accounting for nearly 33 per cent of the gross cultivated area. The area under other cereals had been : wheat 21 per cent. gram 7 per cent, bajra 4 per cent, and maize 2 per cent of the total cropped area in 1944-45.

Cotton is the only non-food crop of importance grown in the state territories. It occupied about 13 percent of the cropped land. Pulses, oilseeds and miscellaneous peasant crops are also grown in the State. These crops include tuar, urad, mung, masur and peas among pulses; tilli, linseed and ground-nut among oilseeds; and poppy, tobacco and sugarcane among miscellaneous crops.

The largest quantity of food crops is grown in the Nimar and Indore districts. In Nemawar the area under food crops is the smallest as compared with other districts. However, the quantity produced is more than sufficient for the needs of the local population, and the food grains used to be exported to other parts of the State earlier. Jowar is grown in the Nimar and Rampura-Bhanpura districts besides Indore and Mahidpur districts. Nimar also produces large quantities of bajra and rice, while the Indore district produced the largest quantity of wheat and gram in the entire state.

Although cotton is cultivated more or less in all the five districts of the state, yet its cultivation is comparatively the largest (about 57 per cent) in Nimar and the remaining 43 per cent being spread over the other four districts. Cotton does not prosper in the deep black soil of Malwa, but with good drainage and manuring the out-turn of cotton in Malwa is being increased year after year. The lighter soils of Nimar, having porous sub-strata, are better suited for cotton cultivation. But the deeper black soil produces cotton of longer staple than the shallow soil; that is why the Malwi cotton fetches better price than the Nimari.[7]

A large number of indigenous plants are cultivated as vegetables in gardens and fields having irrigation facilities. These include many kinds of gourd, cucumber, potato, cauliflower, cabbage, tomato, onion, carrots. yams, garlic, the egg plant,methi, palak etc.

PROBLEMS OF AGRICULTURAL DEVELOPMENT

IRRIGATION : The physical features nd the river system of the State are not favourable to the construction of canal irrigation works, hence the only common source of irrigation in this region are the wells and odhis (wells constructed on the bank of a stream and fed by its water), and tanks. For climatic reasons, however, the proportion of irrigated to inirrigated land in the State is very low. In 1944-45, the irrigated lands covered only 1.3 per cent of the gross cultivated area of the State (Table 5.3).

The crops ordinarily irrigated are poppy, sugarcane, and garden produce. But with the help of irrigation facilities better crops of wheat, maize and barley can be raised in the region.

Table5.3 : Distribution of Irrigated Area
(In thousand ha)

Gross cultivated area	Area irrigated from			Total irrigated area
	Wells	Tanks	Ponds or streams	
2,113	24	2	1	27

Source : Holkar State Administration Report, 1945, Indore Appendix II, pp. 4-5.

MANURES AND FERTILIZERS : Manuring is but little practised in Malwa, being practically confined to fields which bear crops of poppy, sugarcane and vegetables. In Nimar the cultivators, who can afford, manure their fields for the cotton crop every third or fourth year. The common manures are the dung of the cattle and village sweepings. The supply of cow-dung for manuring purpose is not large owing to its extensive use as fuel.

A common method of manuring a field in some tracts is by grazing goats and sheep over it and herding them on it at night. Green manuring is also practised in the State. This process consists of ploughing a growing leguminous crop into the soil before it has reached at maturity. Several plants, such as san, Bombay hemp, kulthi, urad, jawar and chaola are for green-manuring purposes in the State.

Chemical fertilizers particularly ammonium sulphate, urea and super phosphate are being intensively used these days in this region for growing improved crops.

IMPLEMENTS : The implements used in agricultural operations are few and of the simplest kind. In Malwa such implements are of lighter construction than in Nimar, the stiff soil in the latter tract requiring heavier implements. The principal implements are the plough, harrow, knife, sickle, axe, rake and goad for different purposes.

There is need of improved implements for better yield per hectare as 'effective and improved implements will not only save time but bring about efficiency in operations with commensurate returns in the form of increased yields also.'[8] But the large scale mechanization on foreign pattern is unsuited to the country, as 'machinery can be profitably used only to the extent to which it saves labour, that might otherwise be productively employed, or to the extent it performs work that hand-labour can not do, or can not do as well, or can not complete quickly enough to enable farm operations to be done at the most suitable time for maximum production.'[9]

MINERALS

Minerals are significant not only as fundamental requirements of modern industrial society but also because they often figure prominently in territorial dispute. The distribution of minerals in the earth's surface is very uneven.

The territories of the erstwhile Indore state have very few minerals (Figure 14). Iron ore is found in a small quantity at Kethuli and Navali in Rampura-Bhanpura district, and between Bain and Sendrani in the Nemawar district, and between Katkut and Kanar river in Nimar district. Yashwant Rao Holkar I is said to have obtained metal for his gun foundary at Bhanpura from the Kethuli and Navali iron mines, which ceased working in 1884.[10]

Manganese ores are found in the Nemawar district at Bhamar, 11 km north-north-east of Chandgarh. The copper ore appears to have been worked out at Tamkhan in the Nemawar district, where copper veins were 1.2 to 1.5 m wide, extending for about one km. Copper steins were also found between Karia and Jiwani, 5 km to the south-west of Tamkhan.

Among building materials varieties of limestones and sandstones have been worked at places having facilities of transportation. 'Caralline' limestone, which furnishes a handsome marble capable of high polish, was formerly quarried at Bowarla, Kherwan, and Chirakhan in Indore district. It is also available to the west of Barwaha, but here the rock is rather coarse and thin-bedded and has not been worked out. Limestones have been worked at Bhambori and Mori in Rampura-Bhanpura district. Sandstones of excellent quality are also found in the state. Quarries have been worked at Ghatia and Rupabari near Bharwaha, and on extensive scale near Katkut. The stone is durable and soft enough to be carved with facility.

Thus the minerals, because of their scarcity, did not help much in the economy of the old Indore State.

INDUSTRIES

The industrial development of a state is a good index of its power potential. Industrial power means strength for a nation. 'Its importance or lack of it is a great factor in the relations between nations and can be used as a kind of barometre to rank countries.'[11]

With 12.41% of her population dependent on the industries, Indore was an important industrial state among the then princely states of the country. Table 5.4 shows the industrial population of the State in 1941.

COTTAGE INDUSTRIES

Cotton spinning and weaving is carried on in almost all villages. Maheshwar is famous for its handloom industry, which

Table 5.4 : Industrial Population, 1941

	Industry	Actual population dependent upon	Proportion per 10,000 to total population
1.	Textiles	64,808	429
2.	Dress and the toilet	28,961	191
3.	Hides and skins	23,256	153
4.	Wood	17,642	117
5.	Ceramics	9.556	66
6.	Metals	9,556	64
7.	Building industries	7,951	52
8.	Food industries	7,951	52
9.	Chemical products	2,723	18
10.	Miscellaneous industries	18,154	119
	Total	1,87,872	1,241

Source : Census of Central India, 1941, Vl. 1 - Holkar State- Tables, p. 121.

MAP 14

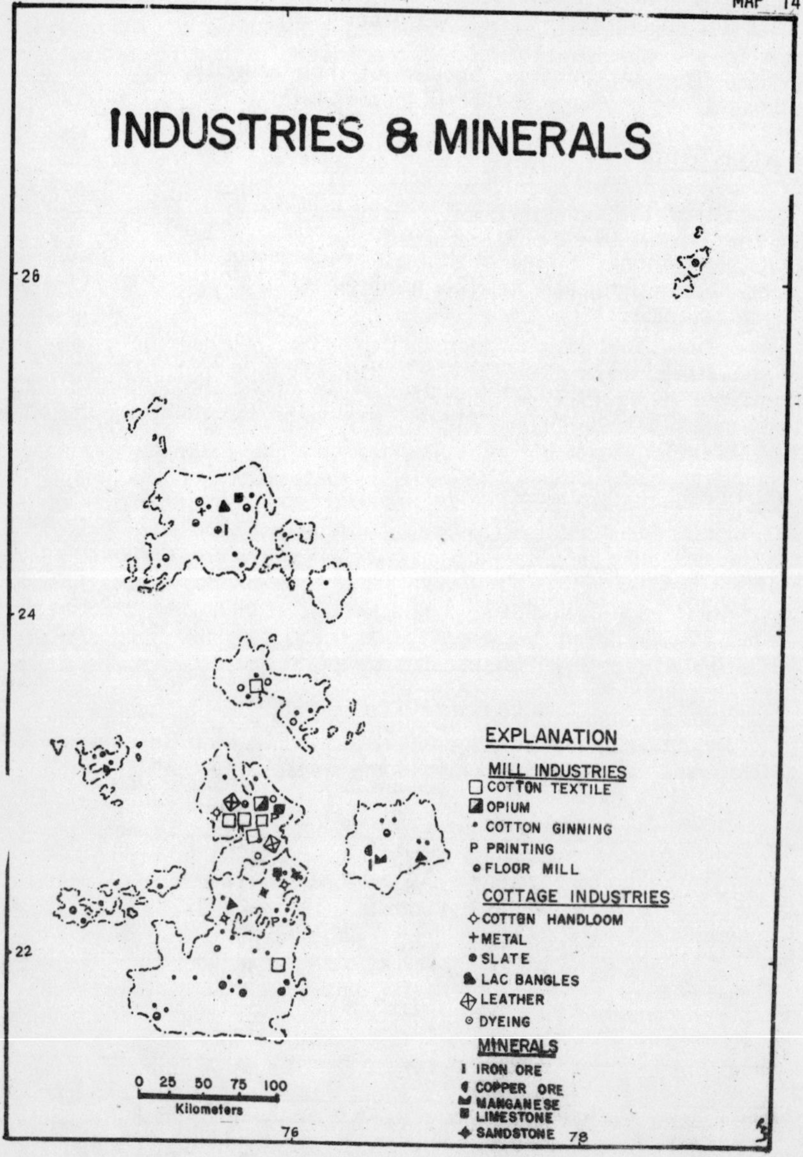
INDUSTRIES & MINERALS
26
24
22
76
78
EXPLANATION
MILL INDUSTRIES
COTTON TEXTILE
OPIUM
COTTON GINNING
P PRINTING
FLOOR MILL
COTTAGE INDUSTRIES
COTTON HANDLOOM
METAL
SLATE
LAC BANGLES
LEATHER
DYEING
MINERALS
IRON ORE
COPPER ORE
MANGANESE
LIMESTONE
SANDSTONE
0 25 50 75 100
Kilometers

specializes in production of the saris and dhotis, dupattas, pagris etc. The Maratha community provides a good market for all these articles. Dyeing and Calico printing are carried on to some extent by the Chhipas[12] at Gautampura, Sawer, Dharwara, Kalmer and Sagwal in Indore district; at Alampur and Mahidpur in Mahidpur district; at Barwaha, Sanawad, Khargone and Gogaon in the Nimar district; and at Rampura and Bolia in the Rampura-Bhanpura district.

Brass, copper and iron utensils and vessels for every day use are manufactured at Parda in the Rampura-Bhanpura district. Khadaoda, in the same district, is famous for her slate industry. Lac bangles are made by the Lakheras[13] in most of the settlements, specially at Maheshwar and Rampura.

Hide tanning and shoe mking is a village industry carried on throughout the State to meet local requirements. Indore and Mhow are important centres of shoe industry. Gur (jaggery) and rab are made from sugarcane juice at Rao and Bijalpur near Indore city. Hand-grinding chakkis are made at Sigonya and and Harangaon in Nemawar district. Most of the crafts and industries, which are strictly urban, are localized at or around Indore city.

MANUFACTURING INDUSTRIES

Maharaja Tukoji Rao Holkar II was the founder of the modern industrial development of the Indore region. The State Mill to manufacture cotton cloth was established with his efforts in 1866.[14] Another textile mill was set up in 1883, and between 1916 and 1927, no less than five other textile mills were started at Indore. At present there are seven spinning and weaving mills at Indore and one at Sanawad, containin a total of 1,57,262 spindles and 4,569 looms, employing around 25,000 persons. These mills turn our coarse cloth, checks, white and khaki drill, dusuti, twills, and other shirtings, alongwith dhotis and saris for male and female wear. The cotton textile industry was the backbone of the state revenue in the Holkar State.

There are a large number of ginning factories all over the region totalling 95, apart from 20 cotton presses. A brass foundry, a knitting and hosiery factory and a paper mill are located at Indore. Dairy, engineering, chemical, small instruments and agricultural implements manufacturing units are also found at and around Indore.

Table 5.5 shows the population of industrial workers according to different industry groups.

The industries of the state, all except the cotton textile industry, suffer from essentially the same handicaps as their counterparts throughout the country. 'Outmoded techniques of production, inefficient organisation, lack of financial resources to buy and sell on the most advantageous terms, dependence on middlemen, narrowness of the market, inability to adopt production to changing conditions of demand - these are oft repeated explanations for the poor state of economic health of small industries.'[15]

Table 5.5 : Industrial Workers in Indore State

	Industry	Total population engaged	Per cent of the industrial labour
1.	Agricultural	15	0.1
2.	Mines and quarries	50	0.2
3.	Textiles	24,998	91.7
4.	Hides and skins	33	0.1
5.	Wood	22	0.1
6.	Metals	165	0.6
7.	Ceramics	527	1.9
8.	Chemical products	89	0.3
9.	Food	250	0.9
10.	Narcoits	95	0.4
11.	Dress and toilet	19	0.1
12.	Transportation construction	1	N
13.	Electrical transmission	212	0.8
14.	Stationery	4	N
15.	Transport	439	1.6
16.	Miscellaneous	310	1.2
	Total	27,276	100.0

Source: Census of Central India, 1941, Vol. I - Holkar State - Tables, p. 129.

After integration in 1948, the city of Indore has seen a rapid development of industries, and now it is the biggest industrial centre of the Central India.

TRADE

Trade is a symptom of civilization. The economic progress of a nation or an individual is based upon trade.[16] Owing to the inland situation of the state, the entire trade was carried on by the land transport, the main trade routes in the State being the Bombay-Agra Road, the Mhow-Neemuch Road, the

Indore-Harda Road, and the Indore-Kukshi-Nisarpur Road. Rail--borne trade in the State was chiefly carried over the Ajmer--Khandwa section of the then Bombay Baroda and Central India Railway (B.B. & C.I.R.), now the Western Railway. Trade in the State provided livelihood to only 2.16 per cent of its population in 1941.

Most of the internal trade of the Holkar State ws carried on in Indore city, the commercial hub of the Central India. Indore is famous for her trade in cloth, gold and silver, cotton and grain, sugar, kerosene oil and metals of all kinds. Senawad, Khargone, Barwaha and Sendhwa, All in the Nimar district, are famous for their trade in cotton and distribution of consumer goods. In other districts the principal markets area at Manasa, Sunel, Garoth, Piplia, Rampura and Bhanpura in the Rampura--Bhanpura district; at Mahidpur, and Tarana in the Mahidpur district; and at Kannod and Khategaon in the Nemawar district. Siyaganj, then at the outskirts of the Indore city, was declared[17] a free zone area in 1891, bringing a boon to the State's trade. At Siyaganj goods were imported free of duty for re-export to foreign territories without any restriction provided they did not enter the State limits.

COMMODITIES OF TRADE

The erstwhile Indore State used to import rice, piece-goods. yarn, salt, sugar, molasses, spices, tobacco, precious and othe metals, kerosene oil, machinery, coal, cotton and foodgrains. The principal exports of the state included cotton, cloth, grain, oilseeds, cattle, opium, spices, hides and skin, timber and tobacco.18

Most of the import trade, thus, included consumers' goods like piece goods, kerosene oil, salt etc. Larger imports of manufactured goods point to the industrial backwardness of the State. Bombay supplied most of the imported commodities in the state.

The exports of the state were getting their way in the neighbouing states of the then Central India and Rajputana Agencies.

MECHANISM OF TRADE

Each district of the state had its recognised collecting and distributing centres, which were fed by the weekly markets held in almost all the villages. The big traders buy their goods through the petty local traders or their own agents, and export them to the chief trade centres of the State-Indore city, Siyaganj, Sanawad, Mhow and Rampura.

For the benefit of the cotton cultivators, the state had opened regular cotton markets at Sanawad, Barwaha and Khargone. The cultivators commanded a competitivee price for their produce through these markets.

Thus, the erstwhile Indore state had a sound economic base behind her comparative socio-poitical superiority among the then Central Indin states, most of them being too small to be economically self-sufficient and politically strong. Her superb economic position has turned the Indore region into the economic core of the Madhya Pradesh, the heartland of India.

References

1. E.H.Carr, The Twenty Years Crisis 1919-1939, London, 1946, p. 140.
2. R.P. Singh Electoral Politics in Manipur:A Spatio-Temporal Study, New Delhi,1981, p.9.
3. Uncultivated land is also called culturable waste land.
4. S.Van Valkenburg and C.L. Stotz, Elements of Political Geography, New Delhi, 1963, p. 63.
5. R.N. Djubey, Economic Geography of India, Allahabad, 1968, p. 187.
6. O.H.K.Spate, India and Pakistan : A General and Regional Geography, London, 1960, p. 253.
7. L.C. Dhariwal, The Indore State Gazetteer, Vol. I, Indore, 1931, p. 139.
8. L.R. singh, The Tarai Region of U.P. - A Study in Human Geography, Allahabad, 1965, p. 86.
9. Charan Singh, Joint Farming X-rayed - The Problem and its Solution, Bombay, 1959, p. 79.
10. Dhariwal, op. cit., p.535.
11. Valkenburg and Stotz, op. cit., p. 197.
12. A Community specializing in dyeing and calico printing. The Chipas are either Hindus or Muslims.
13. A community specializing in manufacture of lac bangles.
14. Dhariwal, op. cit., p. 214.
15. National Council of Applied Economic Research, Techno--Economic Survey of Madhya Pradesh, New Delhi, 1960, p. 72.
16. Dubey, op. cit., p. 218.
18. ibid., p. 217; Imperial Gazetteer of India, Vol. XIII, Oxford, 190[illegible], p. 344.

CHAPTER 6

Circulatory System

Civilization is largely the product of circulation,[1] which denotes the flow of people, armies, material goods, capital, messages and ideas across the space, open to man's activities.[2] The various media through which this flow is made possible form the circulatory system, and include factors as transportation system, radio and the press. It has well been said that 'the roads and railways are the arteries of a country, and the traffic that flows along them is the blood stream that gives it life.'[3]

Modern communications facilitate the movement of commodities, people and ideas. They play very important role in shaping the destiny of the states. In the land-locked State of Indore, roads and railways provided excellent routes of communication.

ROADS

Of all land routes, the road, including everything from the humblest path to the modern concrete highway, is the most ancient as well as the most universl.[4] In Buddihist days a main trade route from Paithana (modern Paithan) to Shravasti (Sahet Mahet in the Gonda district of Uttar Pradesh) traversed this state with a halting stage at Mahissati, now called Maheshwar, the next important stage being Ujeni (Ujjain).[5] In Mughal days the royal road from the Deccan passed through the Nimar District. The halting stages on this road between Bijapur and Ujjain were Bhikangaon, Gogaon, Depalpur and Fatehabd. These early routes represent the marchland position of this region as a whole.

The Indore State had two categories of roads : (1) Government Roads, constructed and maintained by the British Government, and (2) State Roads, constructed and maintained by the Holkar State (Figure 15).

GOVERNMENT ROADS

The oldest metalled road[6] in the region is the Agra-Bombay

MAP 15

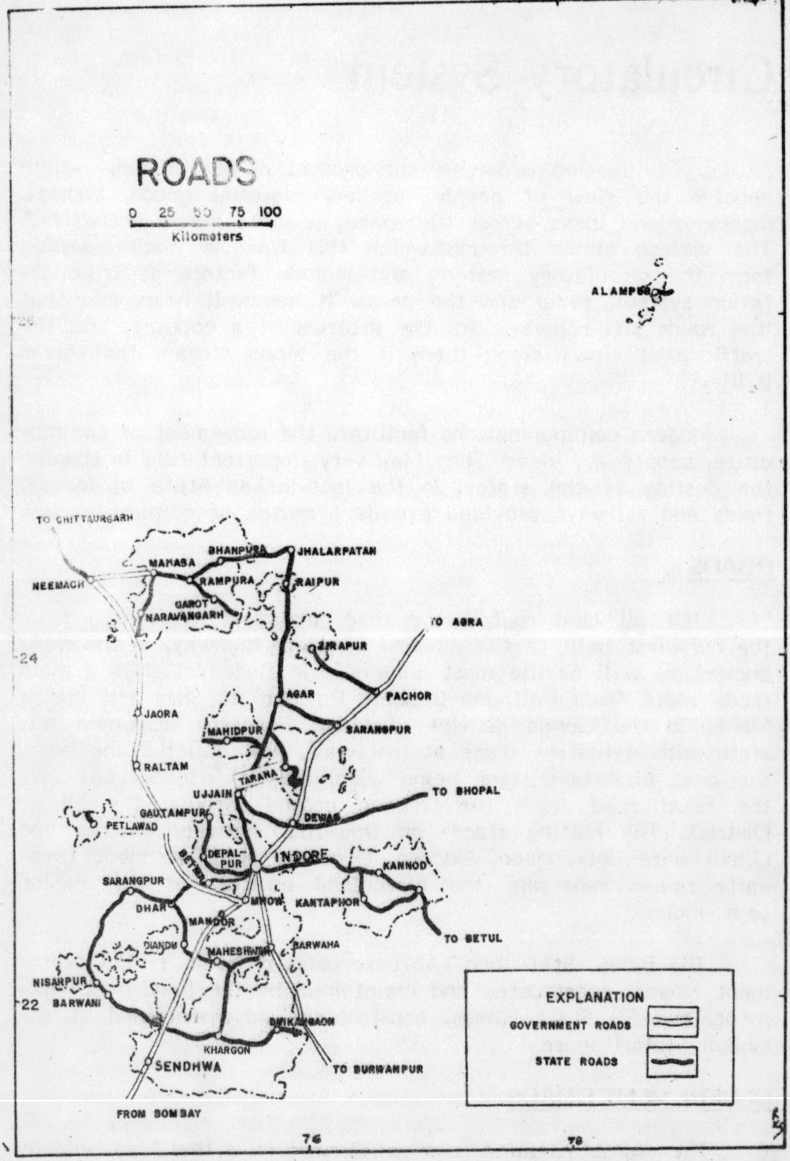
ROADS
0 25 50 75 100
Kilometers
ALAMPUR
TO CHITTAURGARH
NEEMACH
MANASA
BHANPURA
JHALARPATAN
RAMPURA
RAIPUR
GAROT
NARAYANGARH
TO AGRA
PACHOR
AGAR
JAORA
MAHIDPUR
SARANGPUR
RALTAM
TARANA
UJJAIN
TO BHOPAL
GAUTAMPURA
PETLAWAD
DEWAS
DEPAL-PUR
INDORE
BETMA
KANNOD
SARANGPUR
DHAR
MHOW
KANTAPHOR
MAHESHWAR
BARWAHA
TO BETUL
NISARPUR
BARWANI
KHARGON
SENDHWA
TO BURWANPUR
FROM BOMBAY
EXPLANATION
GOVERNMENT ROADS
STATE ROADS
26
24
22
76
78

Road, with a length of about 134 km within the territories of the erstwhile Indore State. Under the Nagpur Plan this road has become the National Highway No.3. Sendhwa, Mhow and Indore are the halting stages on this road in the state limits. Other Government roads have a length of about 121 km. These are the Indore-Simrol-Khandwa Road (80 km), Mhow-Neemuch Road (19 km), Ujjain-Agar Road (19 km), passing through Mahidpur district, and Neemuch-Manasa Road (3 km).

These Government roads formed part of the inter-state road net-work which besides connecting Indian princely states proved to be the lifeline of the British rule in our country. These roads have been of prime militiary importance as they connected not only Mhow cantonment with Indore city but also with other state capitals in the then Central India Agency. Though the Britishers visualized the fundamental unity of the Indian subcontinent and her people, yet they did not construct these roads for the unity and integrity of the country but for the perpetuation of the British rule.

STATE ROADS

Besides the Government roads, the holkar state had constructed and was maintaining about 847 km of the metalled rods connecting various blocks of her territory, district and sub-divisional hedquarters, important town and villages. These state roads, as they were referred to those days, were radiating in different directions from the capital city of Indore or from Khargone, the headquarters of the Nimar district.

State roads connect Indore with Sawer in the north, Depalpur in the north-east, Betwa in the west, and Nemawar in the east through Kannod and Khategaon. A road from Khategaon connects Kantaphod and leads to Dhantalao on Nemawar-Indore Road.

From Khargone roads lead to Daisgaon in the east through Bhikangaon, Sanawad in the north-east, Mandaleshwar through Kasrawad in the north, and Nisarpur via Un, Segaon and Chikhalda in the west. Maheshwar, the former seat of the Holkars, is linked with Barwaha, the nearest railhead on the Western Railway.

Mahidpur is connected by road in the west with Mahidpur Road on the old B.B. & C.I.R., and in the east with Ghosla on Ujjain-Agar Road. A road from Pat on the Ujjain-Agar Road connects Tarana and Makshi on the Bombay-Agra Road.

In Rampura-Bhanpura district the main road links Piplia with Bhanpura through Narayangarh, Manasa and Rampura. Manasa is also linked with Neemuch Cantonment in the west. Other roads in the district connect Rampura with Garoth, Shamgarh with Chandwasa, and Jirapur with Pachor.

In addition to the Government and State roads there are many unmetalled roads connecting important villages and towns to the nearest road or railheads. Under the Nagpur Plan most of these roads have become inter-village roads.

All along these roads the state constructed and maintained a number of dak bangalows and rest houses (Table 6.1). At present most of the towns have roadside facilities for the benefit of the tourists. Of late, the region has become an important area of tourist interest.

Table 6.1: Rest Houses and Dak Bungalows in Indore State

District	No. of Rest Houses and Dak Bungalows
1. Indore	11
2. Mahidpur	6
3. Nemawar	6
4. Nimar	14
5. Rampura-Bhanpua	13

Source : L.C. Dhariwal, The Indore State Gazetteer, Vol.I, Indore, 1931, p. 226.

The erstwhile Indore state, with her 1,102 km of road suitable for motor traffic, had the road-net density of one km to every 23 sq km of her area, or for every 1,374 persons. The road length and the road density had been quite impressive in comparison to most of the princely states of India. Hence, it was possible for the Holkars to integrate most effectively various blocks of territory, which formed part of the Indore State.

RAILWAYS

Railways are relatively modern type of land-route and are not as ancient as the roads. The first rail-route was built in England in 1825, and in 1853 india saw the birth of railways on its soil. The first railway line constructed in the Holkar state was the metre gauge line from Khandwa to Indore, which passed for 99 km through the state's territory. The Holkar State Railway, as it was called those days, was constructed between 1869-77. Subsequently it was pushed on beyond Indore, and was finally linked up with Ajmer via Ratlam. This branch of the then Rajputana-Malwa Railway has a total length of 148 km with 15 stations. Fatehabad-Chandrawatiganj station on this line is linked up with Ujjain. Between Jaora and Neemuch, this line traverses for a few km in the south-western corner of the Rampura-Bhanpura district.

The Ratlam-Godhra broad gauge branch of the former B.B. & C.I.R. passes through Petlawad pargana with a station at Bamnia, while the Nagda-Mathura branch just touched the western fringe of the Mahidpur district, and passes through the south-eastern part of Rampura-Bhanpura district with stations at Shamgarh, Garoth and Kurlasi.

The Ujjain-Bhopal branch of the Great Indian Peninsular Railway (G.I.P.R.), now part of the Central Railway, crosses the south-eastern portion of the Mahidpur district with a station at Tarana Road (Figure 16).

The density of rail-net in the state had been 0.8 km for each 100 sq. km of her area or 13.4 km for 1,00,000 persons,[9] as the State had 203 km of railways within her territory. The railways, which passed through the Holkar state, have become part of the Western Railway after their regrouping in November 1951.

The impact of railways on the economic life of the states has been quite marked. 'Grains can now be carried from one part to another freely, which has largely tended to equalise prices. Railways have also necessiated a relaxation of caste observance while travelling, but on the other hand they have certainly lended to bring members of the same caste living at distance into closer communion.'[10]

AIR ROUTES

Air transport is a development of the past few decades. The particular asset of the air transport is its speed; the handicap, its inability to carry heavy loads or bulk cargo. 'Unlike early rail, motor and water transport, air transport specializes not in local but in long distance carrying.'[11] In Indore state, therefore, air transport was not important though Indore is

MAP 16

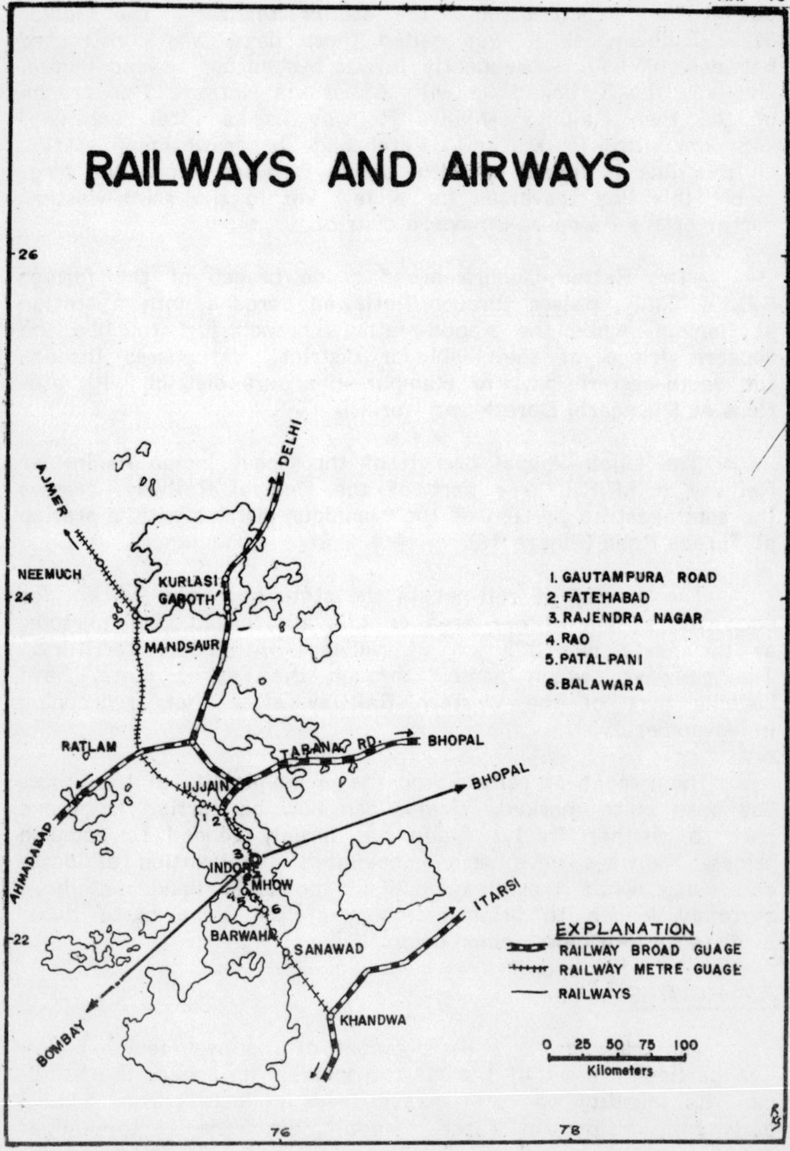
RAILWAYS AND AIRWAYS
26
24
22
76
78
AJMER
DELHI
NEEMUCH
KURLASI
GAROTH
MANDSAUR
RATLAM
UJJAIN
TARANA RD.
BHOPAL
BHOPAL
AHMADABAD
INDORE
MHOW
BARWAHA
SANAWAD
ITARSI
KHANDWA
BOMBAY
1. GAUTAMPURA ROAD
2. FATEHABAD
3. RAJENDRA NAGAR
4. RAO
5. PATALPANI
6. BALAWARA
EXPLANATION
RAILWAY BROAD GUAGE
RAILWAY METRE GUAGE
RAILWAYS
0 25 50 75 100
Kilometers

an important air station in the Central India, on Bombay-Delhi route.

POLITICO-GEOGRAPHIC IMPORTANCE

Transportation facilities have been important in integrating the states' internal economy and in binding them together. The roads and railways of the Holkar state were the most important sources exercising a unifying influence for the emotional integration of its people, who were living in not less than 80 scattered blocks in the Central India. The transport routes connected Mhow and Neemuch contonments with almost all the princely states in the region, and thus were of too much military importance. Government roads and railways were constructed by the British Government visualizing the strategic importance of the region. 'All the military routes to the Deccan passed through it, so also the routes to Gujarat, and to the sea-ports on the western coast. For striking at Rajputana Gondwana or Berar, Malwa afforded an excellent military base.'

The development of transportation facilities probably has beeen the most important forward step of our modern civilization, which has brought the people of different states closer to each other.

References

1. A.E. Moodie, Geography Behind Politics, London, 1963, p. 103.
2. J. Gottman, The Political Partitioning of our World; An Attempt at Analysis, World Politics, Vol. IV. No. 4 (July, 1952), p. 515.
3. E. Hawks, The Romance of Transport, London, 1936, p. 5.
4. V.C.Finch, G.T. Trewartha, A.H. Robinson and E.H. Hammond, Elements of Geography-Physical and Cultural, New York, 1957, p. 626.
5. The Impherial Gazettee of India, Vol. X. Oxford, 198, p.371.
6. L.C. Dhariwal, The Indore State Gazetteer, Vol.1, Indore, 1931, p. 222. Costructed between 1840-1860.
7. Figures in brackets indicate the road length in the territories of the erstwhile Indore State.
8. The Bombay-Thane rail-route was laid in 1853.
9. Dhariwal, op. cit., p. 228.
10. The Imperial Gazetteer of India, op.cit., p. 371.
11. Finch, Trewartha, Robinson and Hammond, op.cit., p. 643.
12. Jadunath Sarkar, History of Aurangzib, Part V, Calcutta, 1952, p.381.

CHAPTER 7

Demographic Pattern

Every state, as Ratzel said, is a piece of land and a piece of humanity.[1] Professor Moodie endorses the Ratzelian view on state when he visualizes state as 'an amalgam of territory, people and the organisational machinery which is set up to guide the relationships between people and their physical environment.'[2] Thus the population is an essential element of the state, without which the state can have neither policy nor the power to carry policy into effect.

The people of any country are important economically both as consumers and as producers,[3] bringing treasures of the economic development to the state. They are the first essential economic resource which any country must have. Population and its associated features-density and distribution patterns, movements and migrations and its areal variations in quality as expressed through such attributes as health, race and nativity, age composition, literacy, religion, occupational status, stage of economic development, and the like influence the politico-geographical behaviour of the states.

DISTRIBUTION AND DENSITY OF POPULATION

The erstwhile Indore State had 15,13,966 persons according to her last census (1941), with a density of 55 persons to the sq km.

In 1820, the population of the State was 5,37,914 as per Mackay's calculations.[4] During the next 120 years it recorded a three-fold increase of 9,76,052 persons. Table 7.1 shows the population of the State enumerated in the decennial censuses beginning from the year 1881.

The data, presented in Table 7.1 shows the population of the State enumerated in the decennial censuses beginning from the year 1881.

The data, presented in Table 7.1. show the continuous growth of the State's population except in 1901, when an unpre-

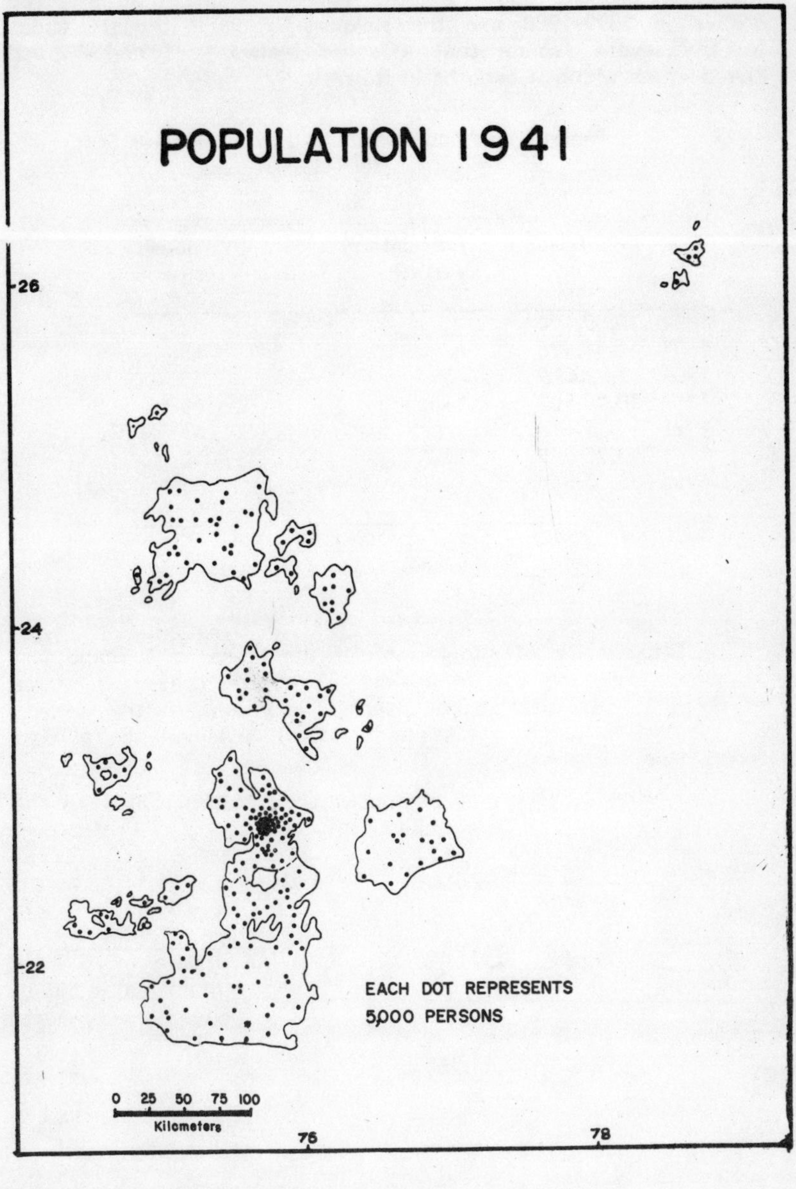
POPULATION 1941
26
24
22
EACH DOT REPRESENTS
5,000 PERSONS
0 25 50 75 100
Kilometers
75
78

cedented drop in population was recorded due to 'the heavy mortality brought about by the severe famine that affected Malwa in 1899-1900 and the sickness which it brought about by the severe famine that affected Malwa in 1899-1900 and the sickness which it brought in its train.'[5]

Table 7.1 : Population Trend in the Holkar State (1881-1941)

Year	Population	Percentage variation	Density per sq km
1881	10,89,154	-	42
1891	11,46,526	4.7	45
1901	9,08,426	-20.8	35
1911	10,58,502	16.6	41
1921	11,55,405	9.2	45
1931	13,25,089	14.7	52
1941	15,13,966	14.3	55

Source : Census of Central India 1941, Vol.1 - Holkar State-Tables, pp. 6-7.

The density of population for the State as a whole was 55 persons per sq km as against 52 at the preceding census of 1931. There were wide internal variations in the density of the administrative units and natural divisions[6] as evident from Table 7.2.

Table 7.2 : Density of Population in Natural Divisions and Districts, 1941

Natural division	Area (km)	Popultion	Density per km	District per km	Density
Plateau	11,588	8,53,748	74	Indore	118
				Mahidpur	64
				Rampura-Bhanpura	44
Hilly	14,027	6,43,360	46	Petlawad	47
				Nemawar	35
				Nimar	49
Plain	114	16,858	148	Alampur	148

Source : Census of Central India 1941, Vol. 1 - Holkar State - Tables, pp. 6-7.

A number of causes operate to bring about such great variations in the density of population in the various parts of the state. The more important amongst them are - the character and configuration of the soil, rainfall, climatic conditions and the state of industrial #velopment. In the areas of favourable geographical conditions the density is generally higher, it being the highest in the low-lying plain area of Alampur, less in the plateau region, and the lowest in the hilly tracts (Figure 18).

A high density of population in the Indore district was due to its industrial development as also due to the location of the seat of the State Government, and political cum cultural 'Capital' of the Central India.[7] Alampur, in the low-lying tract of the Sindh-Pahuj basin, recorded the highest density in the state, which was 'entirely accidental or artificial due to its smallness of its extent, and isolation from rest of the State.'[8]

Whenever the population is mostly agricultural, as it was in this state. there must exist a correlation between the number of people living in a particular area and the proportion of the land actually under cultivation within that area. The regions with greater proportion of cultivated land have always denser population than those with smaller proportion of cultivated area. The climate, the fertility of the soil, the proportion of the area under irrigation, the variety of crops grown, the habits of industry and the standard of living among the agricultural classes, the marketing facilities, the means of communication, the ratio of non-agricultural population and similar other factors disturb the relation between the density and the proportion of cultivated area in different parts of the State.[9]

Table 7.3 shows the density of population and the percentage of gross cultivated and irrigated area, and rainfall for various districts and administrative units in the state.

Nemawar, with a density of 35 persons per sq km, had the most sparse population in the entire state. Its proportion of cultivated area, 53.4 per cent, is somewhat larger than that of Petlawad's 48.3 per cent, and a little lower than that of Rampura-Bhanpura's 65.2 per cent. The irrigated area of Petlawad and Rampura-Bhanpura was obviously responsible for their higher density. Petlawad, for the almost similar density as Rampura-Bhanpura, had lower percentage of cultivated and irrigated area. The primitive Bhils, with a low standard of living, preponderate in Petlawad giving rise to the density of 15 persons to the sq km with a slight difference in their cultivated area, of 5 percent only. This difference in their densities

MAP 18

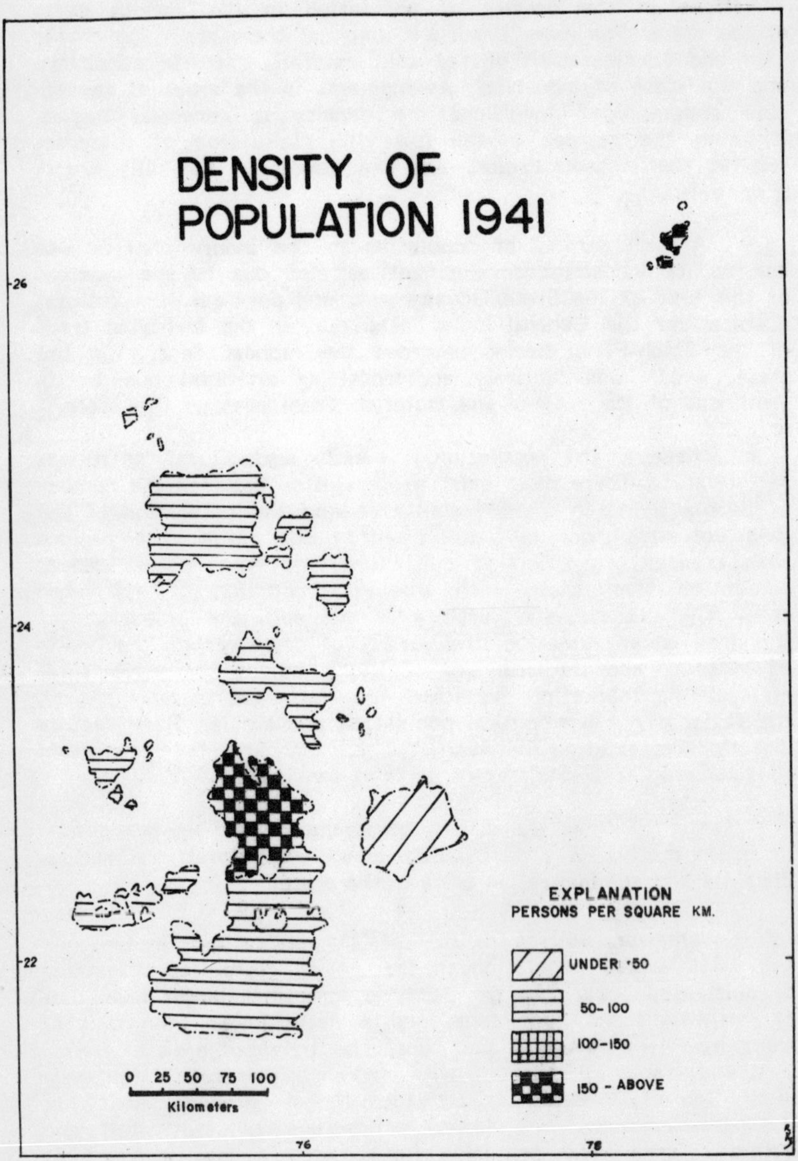
DENSITY OF
POPULATION 1941
26
24
22
76
78
EXPLANATION
PERSONS PER SQUARE KM.
UNDER ·50
50- 100
100-150
150 - ABOVE
0 25 50 75 100
Kilometers

Table 7.3 : Density of Population, Landuse, Water Supply and Crops

District/ Tehsil	Density (Per Km^2)	Percentage of Total Area		Percentage of cultivated area		Percentage of cultivated area Irrigated.	Rain fall (cm)	Percentage of gross cultivated area under				
		cultivable	Net cultivated	Net cultivated	Double Cropped			Wheat	Jowar	Other food grain	Cotton	Other crops
STATE	55	74.9	48.1	64.1	4.3	2.6	71.4	16.8	30.2	18.8	25.1	7.1
Indore City & Mhow Cantt.	5421	-	-	-	-	-	85.6	-	-	-	-	-
Indore (excluding Indore City & Mhow Cantt).	56	89.5	61.6	68.8	2.3	1.6	82.5	48.3	22.8	12.2	7.9	8.8
Mahidpur.	64	84.3	48.6	57.7	5.1	2.6	72.1	20.0	36.6	7.9	29.7	5.8
Rampura Bhanpura	44	63.2	37.2	58.8	6.4	6.8	59.4	13.4	41,8	17.6	17.4	9.8
Petlawad	47	77.3	31.3	40.5	7.8	5.0	94.2	5.3	0.7	55.1	31.0	7.9
Alampur	148	89.2	78.6	88.2	3.8	1.0	45.7	16.8	22.0	54.6	1.1	5.5
Nimar	49	74.7	53.6	71.7	3.9	1.9	62.0	5.9	28.7	26.2	34.8	4.4
Nemawar	35	65.9	32.9	49.9	3.5	0.3	92.7	27.6	24.9	10.3	23.4	13.8

Source : Census of Central India, 1941, Vol-I Holkar State Tables, p.4.

was mainly due to the higher percentage of irrigated area and a higher rainfall in Mahidpur district. Indore had a much higher proportion of cultivated area than Mahidpur district but a lower density.Probably the smaller proportion of the area under cotton in Indore (7.9 per cent) than that of Mahidpur (29.7 per cent) was responsible for this variation. Alampur had the highest density as well as the highest percentage of cultivated area.

The density of population was not too large in the Indore state. After the independence of the country and merger of the state with the Indian Union, the State experienced a faster growth in its population due to 'the advances in medical care and great economic possibilities stemming from better methods of production in both agriculture and manufacturing.'[10]

AGE COMPOSITION

The age structure of the population, the balance between the sexes, the educational attainments and the level of technical efficiency, all are factors in the power potential of the state.

Age composition of population is one of the most significant qualities. It is an index of future population growth, for young populations have higher fertility potentials. It is also an indication of, (a) the actual and potential labour forces in a state, (b) the potential magnitude of the reservoir of military manpower, and (c) the future size of the groups for which educational opportunities must be planned. The stage, which a region or country has reached in the aging process is indicated by the proportion of people in the population who are under 15 or over 65 years of age. In the indore state, 38 percent of the population was under 15 years age group whereas only 2.42 percent of the people were above 65 years of age in 1941(Table 7.4).

The sex balance is also a physical quality of importance. In both war and peace, the age groups by sex indicate the potential military manpower available for wartime service. In 1941 there were 931 females for every 1,000 males, and this number was properly balanced (Table 7.4). The sex ratio indicates a shortage of females, which might be attributed to the fact that during the Second World War, males might have been engaged in active war service away from the state.

RELIGIONS AND CUSTOMS

Religious belief and practices are powerful motivating

Table 7.4: Age Distribution in Indore State
For 10,000 of each sex)

Age group	Males	Females
Under 5	1,323	1,468
5 - 10	1,325	1,295
10 - 15	1,108	1,081
15 - 20	961	995
20 - 25	1,022	1,059
25 - 30	965	947
30 - 35	804	722
35 - 40	641	601
40 - 45	470	443
45 - 50	470	414
50 - 55	309	295
55 - 60	263	263
60- 65	183	187
65 - 70	90	95
Above 70	115	135
Mean age (years)	34.58	34.22

Source : Census of Central India 1941, Vol. I, Holkar State-Tables, p.52.

agents which directly or indirectly influence our mode of living, use of resources, and the degree to which we are able to modify the earth's surface. 'Religion also exerts a powerful unifying force in so far as the shared experience of its members give rise to a feeling of "belonging together".'[12]

The Holkar state was essentially a Hindu state. Its homogenous religious distribution was appreciably affected only by the presence of Muslims. Christians, Jains, Sikhs, Parsis, Jews and Buddhists were numerically of not much importance (Table 7.5).

The Hindus dominated the scene with their overwhelming majority, amounting to 89.91 per cent of the total population. The Hindus outnumbered the adherents of all other creeds by nine to one. There were 11 Hindus to every Muslim, over 72 to every Jain, and over 264 to every Christian in the state.

Muslims are more of an urban, rather than rural, community. Jains, too, like Muslims, are an urban community and are engaged in trade and commerce. Most of other religious communities are mainly urban.

Table 7.5 : Population According to Religions

Religion	Population	Perecentage
Hindus	13,61,129	89.91
Muslims	1,26,187	8.34
Jains	18,778	1.24
Christians	5,176	0.34
Sikhs	1,870	0.12
Parsis	770	0.05
Jews	55	N
Buddhist	1	N

Source: Census of Central India 1941, Vol. I, Holkar State- Tables, p. 159.

Table 7.6 : Religion-wise Distribution of Population in Administrative Divisions

Religion	Indore	Mahidpur	Rampura-Bhanpura	Nimar	Nemawar
Hindus	3,70,702	2,17,425	1,90,862	4,94,310	87,830
Muslims	64,257	11,252	14,101	27,464	9.113
Christians	4,562	146	47	408	13
Sikhs	1,432	46	31	291	70
Parsis	755	-	6	9	-
Jains	9,157	2,153	3,705	2,717	1,046
Jews	55	-	-	-	-
Buddhist	1	-	-	-	-

Source : Census of Central India 1941, Vol. I, Holkar State- Tables, PP.160-61.

The Muslims numbered 8.34 per cent of the total population. The Jains formed 1.24 per cent of the population. The Christians and the followers of other religions were so few that they hardly made 0.5 per cent of the total population of the state. Except the Hindus, no other caste or creed could, therefore, effect any influence on the working of the State.

The population of Hindus was, however, divided into a number of sub-castes (Table 7.7).

The tribes among the Hindus include : 1. Bhils (including Mankar, Nayak, Patlia, Tarvi, etc.), 2. Bhilala, 3. Barela. 4. Nihal, 5. Gond and 6. Korbu.[15]

Table 7.7: Sub-castes of Hindus

Sub-caste	No. of persons
Harijans	2,37,919
Tribes	2,82,602
Arya Samajis	2,895
Brahmos	41
Other Hindus	8,37,672
TOTAL	13,61,129

Source : Census of Central India 1941, Vol. I-Holkar State-Tables, pp. 159-61.

The term Harijan has been used in the state as it obtained official recognition there and has the same significance as the term "Scheduled Castes" in the post-independence Indian census. The harijans include : 1. Bagri, 2. Balai, 3. Bansphor, 4. Bargunda, 5. Bhambi, 6. Bhangi, 7. Chamar (mochi), 8. Jhamral, 9. Khatik, 10. Koli or Kori, 11. koshti, 12. Mahar, 13. Mang, 14. Pardhi, and 15. Pasi.[14]

Religion has ceased to be an important factor in national unity over much of the world.[15] As such, Indore State never recognised a 'State religion' and remained secular in traditions, folkways, feasts and festivals till its integration with the Indian Union. The Holkars, who originally belonged to Maharashtra, had brought with them not only bravery of the marathas but also many of the festivals and customs, full of patriotism and love for motherland to the Central India.

LANGUAGES

Language is the vehicle of thought and the medium of social intercourse. 'Speakers of different languages possess somewhat different mental make-up, and their thought processes do not run along exactly the same lines.'[16] Therefore, in a state, where majority of the people speak the same language, achievement of emotional unity is easier.

The languages of the State fall under five families as per Linguistic Survey of India, leaving aside the unclassified Gipsy languages. The number of speakers under each of these families is shown in Table 7.8.

The Indo-European language family, thus, claimed 99 per cent of the speakers in the state.

Table 7.8 : Language Families and their Speakers in Indore State, 1941

Language family	No. of Speakers
All Languages	15,13,966
Indo-European	15,00,439
Dravidian	10,406
Austric	1,808
Tibeto-Chinese	6
Semitic	27
Gipsy	1,280

Source :Census of Central India 1941, Vol. I-Holkar State- Tables, pp. 150-53.

There was plurality of languages and dialects in the Indore state. The census of 1941 recorded no less than 41 languages in the state (Table 7.9). Of these, four had only one speaker each, other four had less than 10 speakers each. The number of persons speaking 13 other languages varied from 10 to 49. On the other hand, first five languages, i.e. Western Hindi, Rajasthani, Bhili, Marathi and Gujarati, accounted for about 98 per cent of the state population (Bar graph in Figure 19). They all belong to the Aryan sub-family of the Indo-Aryan branch of Indo-European family of languages.

Among the first five languages, which may be treated as the major languages of the state, Western Hindi and Rajasthani are the varients of Hindi - a composite language of the Madhya Desh and the national language of India.

Table 7.10 presents the number of persons speaking Hindi in the erstwhile Indore state and in her various administrative and census divisions and the proportion they bear to the population.

The state had registered about four-fifths of its population as Hindi-speaking in 1941 (Figure 19). Hindi, being almost universal in Alampur and Mahidpur-Tarana, was spoken by 97.71 per cent of people in Jirapur, by 96.87 per cent in Rampura-Bhanpura district and by 96.52 per cent in Indore district (excluding the city and cantonment areas). Nemawar had 91.96 per cent of its population as Hindi-speaking. The cosmopolitan population in Indore City and Mhow Cantonment had 81.32 and 80.83 per cent of their people respectively as Hindi-speaking. In Nimar and Petlawad the lower proportion of Hindi-speakers was mainly

Table 7.9 : Indore State - Languages at a glance

Language	No. of Speakers	Language	No. of Speakers
Western Hindi	6,40,054	Rajasthani	5,70,807
Bhili	1,98,266	Marathi	48,714
Gujarati	28,377	Gondi	8,388
Eastern Hindi	4,836	Punjabi	2,826
Khandeshi	3,071	Nihali	1,617
English	1,509	Gypsi	1,280
Tamil	1,221	Sindhi	1,079
Telugu	632	Bengali	325
Pashto	188	Korku	155
Konkani	122	Kanarese	111
Persian	49	Malyalam	47
Eastern Pahari (Nepali)	45	Balochi	34
Kherwari	36	Kashmiri	29
Central Pahari(Garhwali)	30	Arabic	27
Portugese	22	Lahanda	19
French	16	German	12
Oriya	10	Koorgi	7
Chinese	5	Bihari	4
Spanish	2	Burmese	1
Scottish	1	Yugolsavian	1
Hungarian	1		

Source: Census of Central India 1941, Vol. I - Holkar State-Tables, pp. 150-53.

Table 7.10: Hindi Speaking Population in Indore State

Administrative and census division	Population, 1941		Percentage of speakers to total population
	Total	Hindi speking	
STATE	15,13,966	12,15,697	80.29
Indore District	4,50,921	4,00,233	88.76
Indore City	2,03,695	1,65,653	81.32
Mhow Cantonment	34,823	28,153	80.83
Indore District	2,12,403	2,06,427	96.52
Mahidpur District	2,31,022	2,15,594	93.32
Mahidpur-Tarana	1,42.337	1,41,412	99.35
Jirapur	51,738	51,556	97.71
Petlawad	20,089	5,794	28.84
Alampur	16,858	16,838	99.84
Rampura-Bhanpura Dist.	2,08,752	2,02,127	96.87
Nimar District	5,25,199	3,08,651	58.74
Nemawar District	98,072	89,192	91.96

Source :Census of Central India 1941, Vol. I-Holkar State-Tables, pp. 150-53

MAP 19

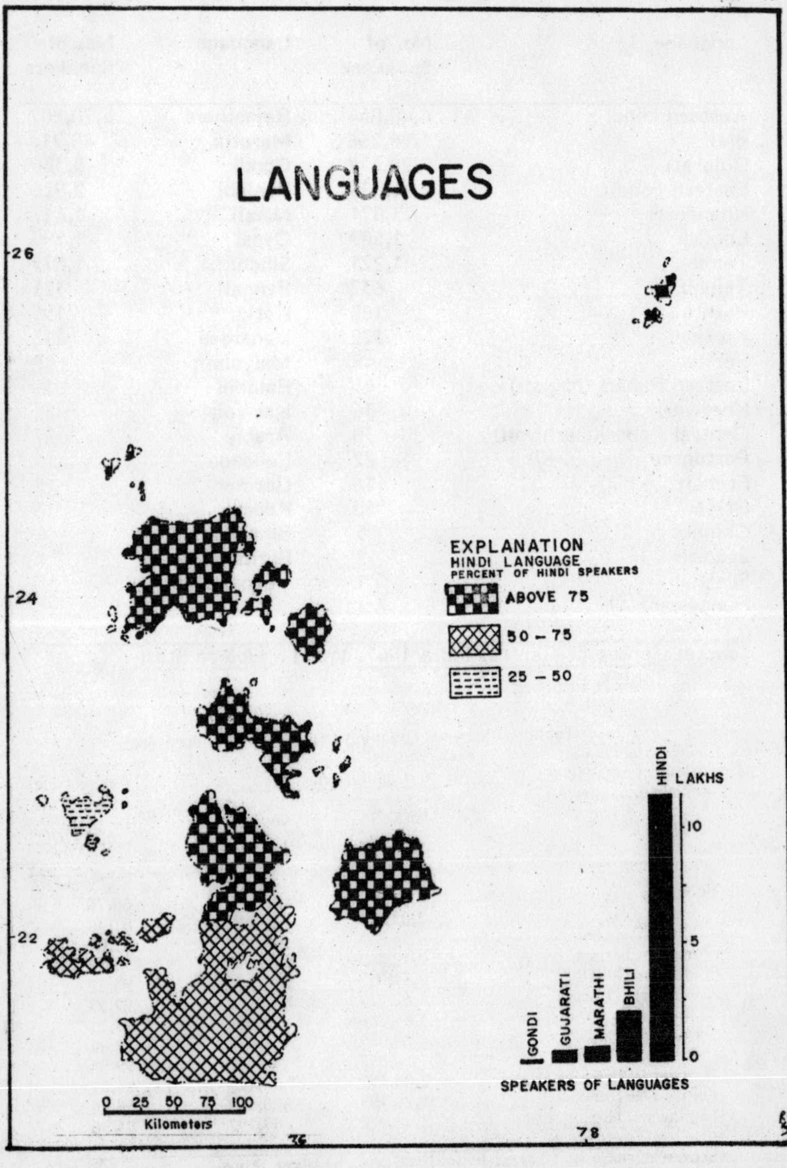
LANGUAGES
26
24
22
EXPLANATION
HINDI LANGUAGE
PERCENT OF HINDI SPEAKERS
ABOVE 75
50 – 75
25 – 50
HINDI
LAKHS
10
5
0
GONDI
GUJARATI
MARATHI
BHILI
SPEAKERS OF LANGUAGES
0 25 50 75 100
Kilometers
78

due to their tribal population who speak Bhili. The percentage of population speaking Hindi in Nimar was 58.74 and in Petlawad 28.84. More and more Bhils are studying Hindi these days, and after some time these areas may completely become Hindi--speaking, as the Bhili language has no literature of its own and Hindi is the only language being taught in the schools.

Bhili, the language of primitive tribes, has its strong-hold in the Nimar district and Petlawad pargana, and accounted for 13.09 per cent of state population. Marathi speakers are mostly confined to the administrative headquarters of the districts and parganas, 62 per cent of them in Indore City alone. Due to proximity to Maharashtra, they are also found in Nimar district in fairly large numbers. Bohras in Rampura-Bhanpura district and Indore City and Mhow Cantonment speak Gujarati.

BI-LINGUALISM

Bi- and even tri-lingualism have been wide spread among many classes of the society except the linguistically homogenous rural areas.Table 7.11 shows the number of people speaking languages selected first as mother-tongue and then as a subsidiary language in the State.

Table 7.11 : Bi-lingualism in Indore State

Language	Speaking as mother tongue	Speaking as subsidiary languge
Hindi	12,15,697	1,12,291
Bhili	1,98,266	1,699
Marathi	48,714	1,857
Gujarati	28,377	1,921
Gondi	8,388	442

Source : Census of Central India 1941, Vol. I-Holkar State-Tables, pp. 154-58.

The number of those who speak languages other than Hindi as subsidiary langauge[17] was very small.

Table 7.12 depects the extent to which Hindi was used as subsidiary language by speakers of Bhili, Marathi, Gujaraati and Gondi.

Thus, 24 per cent of the persons speaking Bhili were speaking Hindi as subsidiary language. This proportion was very low

Table 7.12: Persons Speaking Hindi as Subsidiary Language

Language	Speakers of language as mother-tongue	Speaking Hindi as a subsidiary language	Percentage of those using Hindi as subsidiary language
Bhili	1,98,266	46,915	24
Marathi	48,714	31,698	65
Gujarati	28,377	18,007	64
Gondi	8,388	7,633	91

Source : Census of Central India 1941, Vol. I -Holkar State-Tables, p. 154.

In comparison to the persons speaking Hindi as subsidiary language by the people of other language groups, for the form of Bhili used in the State was quite similar to the local Rajasthani dialects and, therefore, there was a little need to adopt a subsidiary language. Gondi, as a tongue, is entirely dissimilar to Hindi; speakers of Gondi, therefore, are forced to use Hindi for their conversation with outsiders.

Of the persons speaking Marathi, 65 percent had been returned as speaking Hindi as subsidiary language, but it seems to be an understatement. Same was the case with Gujarati speaking population. Most of the persons speaking Marathi and Gujarati had become bi-lingual and learnt to speak Hindi from their very childhood alongwith their mother-tongue.

Before 1901, Marathi, the language of the ruling family was the official language and Modi the official script of the state. But to suit the popular sentiments of the people they were replaced by Hindi language and Devnagari script.[18]

One language can unite the people. Two languages are sure to divide.[19] In the Holkar state Hindi served as the second important centrepetal force for emotional integration of the people living in a number of blocks of territory alongwith the common rule. It saved the state from a 'stratification of society into linguistic layers, each of which would have been diverced from full assoication with its fellows even they might have been equal in law.'[20]

EDUCATION

Educational status, without doubt, is one of the best indicators of population quality. The amount and quality of schooling received by children are a measure of the extent to which society is investing in the well-being of future generations.[21]

The proportion of literate persons in the state in 1941 was 14.65 per cent as against 13.07 per cent[22] for the country as a whole. The male literacy of 23.32 per cent had been much larger than that of the female literacy of 5.19 per cent. It implies that one male in every five, and one female in every twenty was able to read and write. In other words there was one literate female to every four literate males.

The extent of literacy in a particular area depends upon its urbanization, the social structure of its population and facilities for education. In town occupations requiring a knowledge of letters create a demand for literacy; and the higher classes generally live in towns in greater proportion on account of the greater importance of towns in the social, economic and political life. Arts, handicrafts, trade, commerce, and industry offer a better field for activity to men of superior intelligence and training. The administrative headquarters are always in the towns. All these factors contribute to raise literacy in the towns. On the other hand, the rural areas are entirely agricultural, where poor economic conditions and primitive modes of agriculture hardly tend to create any need for reading and writing. The Indian caste system had assigned occupations to groups of poeple; those who followed occupations requiring a knowledge of letters had to be literate, while the others remained illiterate. The large communities, low in the social scale, such as the Balais and Chamars, have always remained outside the pale of literacy. The Arya Samaj and the Christian missionaries could not do well in this field as they are treated intruders by the people, who seem to content with their present miserable lot.[23] Under such circumstances literacy could not make much headway in the state.

Table 7.13 shows the distribution of literacy in various census divisions. Indore City and Mhow Cantonment differed almost in every respect from rest of the state, and their inclusion in the state totals make the entire statistics rather unrealistic. The city and cantonment naturally show a high proportion of literates, about three times of the state average, while all the other divisions stand far below them. The most backward part of the state was Petlawad alongwith Mahidpur, Nemawar and Nimar, where literacy was low (below 10 per cent). Indore and Rampura-Bhanpura districts and Alampur pargana had the medium literacy (11-14 per cent), closer to the state average.

The population expansion, like any region in the world,

Table 7.13: Literacy in Indore State

District and Census Division	Population (1941)			Literacy (%)		
	P	M	F	P	M	F
STATE	15,13,966	7,84,134	7,29,832	14.65	23.32	5.19
Indore City	2,03,695	1,15,298	88,397	38.06	51,58	21.80
Mhow Cantonment	34,823	19,693	15,130	39.11	51.58	22.51
Indore District	2,12,403	1,07,389	1,05,014	11.44	20.20	2.35
Mahidpur	1,94,075	99,448	94,627	7.47	12.94	1.65
Petlawad	20,089	10,167	9,922	6.93	11.94	1.80
Alampur	16,858	8,665	8,193	13.80	24.63	2.28
Rampura-Bhanpura	2,08,752	1,06,359	1,02,393	12.84	21.64	3.77
Nimar	5,25,199	2,67,033	2,58,166	9.76	16.90	2.28
Nemawar	98,072	50,087	47,990	8.84	15.31	2.06

Source : Census of Central India 1941, Vol. I - Holkar State-Tables, pp. 2.147.

may cause serious geopolitical consequences in future in the Central India too, which accommodate the territories of the erstwhile Indore State, because of spatial and economic limitations. A great care of available resources and their careful planning must follow the expanding population for a better standard of living and an all-round development of the people.

References

1. Friedrich Ratzel, Politische Geographie, Munich, 1897, p.2.
2. A.E.Moodie, Geography Behind Politics, London 1963, p.133.
3. P. Fordham, The Geography of African Affairs, Middlesex, 1965, p. 73.
4. A. Mackay, The Chiefs of the Central India, as quoted in L.C. Dhariwal, The Indore State Gazetter, Vol. I, Indore, p.61.
5. Dhariwal, op. cit., p. 63.
6. Natural divisions in this chapter have little relation with the physiographic divisions described in Chapter 4. They have been adopted from the census classification as given in the Census of Central India 1941, Vol.I-Holkar State.
7. Indore was the headquarters of the Central India Agency and Agent to the Governor General used to reside there.
8. Central India Census Series 1921, Vol.1-Holkar State, Indore, 1921, p. 8.

9. Census of Central India 1931, Vol. I-Holkar State, Indore, 1931, p.15.
10. S. Van Valkenburg and C.L. Stotz, Elements of political Geography, New Delhi, 1963, p. 308.
11. V.C.Finch, G.T.Trewartha, A.H. Robinson and F.H. Hammond, Elements of Geography-Physical and Cultural, New York, pp.531-32.
12. Moodie, op. cit., p. 53.
13. Surendranath Dube, Census of Central India 1941, Vol. I-Holkar State- Tables, Indore, 1941, p. 159.
14. ibid., p.159.
15. Norman J.G. Pounds, Political Geography, New York, 1963 p. 9.
16. S. Rundle, Language as a social and Political Factor in Europe, London, 1946, Chapter III.
17. Subsidiary language is the language in common use outside the home.
18. Census of Central India Series 1921, op. cit., p.129.
19. B.R. Ambedkar, Thoughts on Linguistic States, Delhi, 1955, p. 12.
20. Moodie, op.cit., p. 53.
21. Finch, Trewartha, Robinson and Hammond, op. cit., p. 534.
22. Projected figure for all-India literacy.
23. Census of Central India 1931, op. cit., p. 201.

CHAPTER 8

Social Structure

The political structure of a state rests on the fabric of the social structure. Indeed the social structure has a profound impact on the political framework. 'It may modify it in its working. It may nullify it or it may even make a mockery of it.'[1]

Like rest of India, in Indore state too, the social structure was built up on the caste system, the special product of the Hindu civilization and culture. A caste is 'an endogamous group or a collection of several groups, bearing a common name, who by reason of similarity of traditional occupations and reputed origin, are generally regarded by those of their castemen, who are competent to give an opinion as forming a single homogeneous community, the constituent parts of which are more nearly related to each other than they are to any other section of the society.'[2] Thus, endogamy, traditional occupation and a common origin are the three principles on which caste is based.

TYPE OF CASTES

A number of castes inhabit the Central India in general and Malwa in particular. The castes of the erstwhile Indore state fall under following four categories:

1. Functional, such as Lohar, Sutar, Darji etc.
2. Sectarian, such as Gusain, Jogi etc.
3. Tribal, such as Bhils, Gonds etc. and
4. Mixed, such as Khawas, Kunwar, Vidur etc.

Caste at first sight appears fixed and immutable but many new castes have been formed by separate settlements. Being discontented with their present stock, they form new habits and customs which alienate them. Besides settlement, there are a few other powerful factors which greatly influence and help the formation of new castes. They are-

i. Cleanliness or dirtiness of profession;

ii. The acceptance of marriage within or without the caste or of widow marrige; and

iii. Strictness in the exclusions of the forbidden food.

Among Hindus the sub-castes are due more to fission, i.e. separation from the main castes than to acceretion from other castes. Among Muslims the castes and sub-castes formed after the Hindu system as an effect of accretion from Hindu converts and not due to fission among themselves.

CASTE RESTRICTIONS

A number of hard restrictions have been imposed by the Hindu laws and customs with respect to the life of a Hindu. Although nowadays many of them are not strictly observed, the following may be mentioned as nominally binding on every person in the caste

1. A Hindu should marry within his own sub-caste and he should avoid his own gotra or got.
2. Every Hindu girl should be married before she attains puberty.
3. A widow of a Brahmin, Rajput and Bania should not re--marry. Those who practise this are generally considered degraded, and no strict person in the caste should hold communication with them; inter-marry, eat or drink with them.
4. That a man should not take food or water from the hands of people from prohibited castes and castes lower than himself.
5. That he should not follow an occupation derogatory to his caste. This is honoured more in breach than in following.
6. That Mahajans and certain class of Brahmins are prohibited from eating flesh and fish or drinking wine or fermented liquors.
7. They should not touch the untouchables. And if they over do so they must purify themselves by bathing and washing their clothes.
8. They should not draw water from a well used by untouchable classes or allow them to draw from their own wells.

CASTE COMPOSITION IN INDORE STATE

There was considerable diversity in the caste composition of the state due to its widely scattered territories. The typical castes of Malwa inhabit the central part of the erstwhile state, while the northern areas have a number of castes common to Rajasthan. Nimar accomodates a large number of primitive tribes and castes common to neighbouring Gujarat and Maharashtra districts.

The castes having more than one per cent of the total population in the State are shown in Table 8.1. Of the 25 castes, listed there, only 12 claimed more than two per cent, and only six (inclusive of other Muslims), more than five per cent.

Table 8.1: Caste-wise Population in Indore State

Major Caste	Population	Major Caste	Population
Balai	1,08,778	Bhil	1,02,676
Brahmin	94,148	Bhilala	93,682
Rajput	92,136	Other Muslims	90,472
Chamar	73,473	Barela	53,201
Gujar	51,614	Kulmi	45,234
Bania	39,348	Ahir	33,338
Sondhia	31,230	Banjara	27,162
Khati	26,032	Mali	22,460
Deshwali	21,018	Gadaria	20,134
Babaji	18,822	Kumhar	18,753
Nai	17,859	Teli	17,651
Kalota	17,445	Sutar	16,616
Bhoi	15,812		

Source : Census of Central India 1941, Vol. I-Holkar State- Tables, pp. 224-51.

The Balais, with a population of 1,08,778 persons, form the largest caste in the state. This place of pride would have gone to the Bhils, had not the Barelas emerged from their ranks as a separate caste. The Bhils come next to the Balais, and followed by the Brahmins and the Bhilalas. The Rajputs stood fifth. They would have ranked lower if other castes with doubtful claims had not entered in the Rajput returns. The Chamars formed nearly five per cent of the state population, and alongwith Bhilalas and Barelas, who have sprung mainly from the Bhil stock, claimed over one-seventh of the total population.

CASTE GROUPS

Various castes living in the state territories may conveniently be grouped under the following categories-

1. Advanced;
2. Intermediate;
3. Depressed; and
4. Primitive.

The advanced group includes Brahmins, Banias, Kayasthas, Boharas, Parsis, European and Anglo-Indians. This group contains the most advanced section of population, which stands for above the rest of the inhabitants. The group depressed class consists of castes which labour under certain disabilities. The primitive group consists of the Bhils and allied tribes, and Gonds, Korkus and Nihals. The remaining castes, which rank neither very high nor very low, have been termed intermediate.

DEPRESSED CLASSES

The term 'depressed classes' refers to that portion of the Hindu population, which labours under certain disabilities attached to certain occupational castes. Members of certain castes may not enter Hindu temples or may not touch high caste Hindus or approach them within a certain distance, and thus cause pollution; they may not draw water from common well, or reside in localities where high caste Hindus live. Although modern life hs made it difficult for those restrictions to be observed, yet a breach of the established custom is suffered only to the extent that it cannot be avoided. A Brahmin or a Bania cannot object to a Bhangi travelling in the same compartment of a railway carriage, but he would never allow a Bhangi to enter his house.

It is extremely difficult to define the limits of an oppressed class in the ranks of Hindus. Between the highest and the lowest, there is a wide range and variety of castes. In the gigantic hierarchy of caste the social status of a particular caste differs only imperceptibly from that of the one immediately above or below it. It, therefore, becomes impossible to have a very definite line of division between the lower and higher castes or between the touchable and untouchable castes. That the Brahmin and the Bania are among the highest and the Chamar and the Bhangi among the lowest is all too obvious. But it is difficult to decide whether Sunar should rank among the higher castess or not, or whether Telis should be regarded as untouchable. Sunars have now begun to wear sacred thread and the moneyed and educated among them rightly or wrongly believe that they are Kshattriyas, similarly Telis are anxious to pass as Banias. Some may honestly support their claim, while others may honestly reject them. It is really difficult to decide such questions.

The social status of a caste differs spatially. In Uttar Pradesh the higher castes would never drink water from the hands of a Kumhar, but in Malwa the Kumhar is considered just the proper water-bearer. A Deccani Brahmin will refuse to take water from the hands of a Nai (barber), but a Kanya--kubja Brahmin regards him particularly suitable for such service. Again the social status of a caste keeps on varying temporally too. The Kalalas, who are no longer distillers, have risen in status. In the recent past, a number of intermediate castes have declared themselves an Kshattriyas, while the Lohar and Sutar have become Brahmins.

It is difficult to ascertain the exact status of a caste owing, on the one hand, to the anxiety of its members to show it as something higher, and on the other, to the class snobbery of members of higher castes in regarding it as something lower than it really is.

UNTOUCHABILITY

Untouchability can be regarded as the primary test for labelling a caste depressed or otherwise. Bhangi is the only caste in the Malwa region, which is everywhere regarded as untouchable. Chamars and Balais may be considered untouchable but on many occasions they come into actual contact with higher castes without necessrily causing pollution. The more orthodo9 may always avoid them, but generally actual contact within certain limits is not resented. A Brahmin Patwari (village record-keeped) is not anxious to avoid the Balai's touch, when he happens to survey a field and the latter has to carry the chain. In course of agricultural operations, the Rajput has many a time to work shoulder to shoulder with the Balai.

Untouchability in its pronounced form has never existed in Malwa. Malcolm a century and a half ago remarked that Brahminism in its puritanic form was never thoroughly established here. While the most humilitating forms of untouchability, such as those by sight or approach, have never been heard of in this part of India. Only Bhangis were untouchables in stricter sense. The Balais, Chamars and others of similar status have always suffered from disabilities. They are denied entry into Hindu temples, the use of common wells, and residence within the village habitation of higher castes.[3]

The total population of depressed castes in the Indore State was 2,37,919 in 1941,[4] which constituted 15.7 per cent of the total population. In the Holkar State the depressed classes had been termed 'Harijans', a term equivalent to 'scheduled castes' in the Indian Constitution. The castes included under Harijans alongwith their population and characteristics are shown below (Table 8.2).

Table 8.2: Scheduled Castes in Indore State

Caste	Population (1941)	Characteristics
Bagri	11,964	A semi-primitive tribe with a reputation for crimes; generally treated as low and impure.
Balai	1,08,778	Considered to be akin to the Chamars; generally treated as untouchable.
Bansphor	1,291	Workers in bamboo; regarded as untouchable.
Bargunda	1,286	A vagrant tribe of basket and mat makers; treated as untouchable.
Bhambi	3,160	Weavers; regardedas akin to Balais.
Bhangi	9,400	Scavengers, absolutely untouchable.
Chamar	73,473	Skimmers of dead animals and workers in leather; untouchable.
Kasai, Khatik and Butchers	2,840	Regarded as untouchable.
Koli	13,996	A semi-primitive tribe of very low status.
Koshti	2,318	Weavers; considered untouchable.
Mahar	2,574	Village drudges from the Deccan, where strictly untouchable.
Mang	1,243	Makers of mats and brooms; treated as untouchable
Pardhi	1.265	Hunters and fowlers; untouchable.
Pasi	2,556	Breeders of pigs; untouchable.
Others	1,775	Generally untouchables.

UNCLEAN OCCUPATION NOT A TEST OF UNTOUCHABILITY

The occupations of only some of the castes listed in Table 8.2 are unclean. The Bhangi, Chamar, Kasai and Pardhi follow occupations, which from the Hindu point of view, are regarded as unclean. For some inexplicable reason the weaving of cloth or making of baskets, mats and brooms are deemed to be unclean occupations. Bhambis, Koris and Koshtis are considered untouchables for the simple reason that they are weavers, and Bahsphors, Bargundas and Mangs are similarly treated because they make baskets etc. The condemnation of these occupations unclean leads one to conclude that in their case it is not the occupation that makes the caste unclean but the caste that makes the occupation unclean.

The depressed classes represent that section of the pre--Aryan population of India which was not absorbed in the regular Aryan communities. The subjugated people were treated as a servile class, which was kept scruplously apart and made to undergo humilities that weakened them for all time to come. Only the lower occupations were left open to them, and they were made to do the work that the Aryans did not like to do themselves. It has generally been accepted that the Aryans brought with them a higher civilization, or that they developed a higher civilization in India than the one existed here before their arrival. In such a case, the making of baskets and mats out of cane and leaves, a primitive occupation, would naturally be left to a backward and subjugated people. Weaving of cloth, a dull occupation, would similarly fall to their lot. As the conquerors would not stoop to follow such occupations, they would naturally, in the course of time, come to be regarded as degraded, and in the characteristic Hindu way, impure.

Thus the theory that untouchability had its roots in the uncleanliness of occupations appears to be without substance. Centuries must have passed since a considerable section of the Chamars gave up the skimming of dead animals, or even working in leather, and took to agriculture, but there is no appreciable lessening of the rigour of untouchability in their case.

THE PRIMITIVE TRIBES

The erstwhile Indore state had a large proportion of tribal population. In 1941, the primitive tribes numbered 2,82,602 and formed 18.66 per cent of the state population. The primitive tribes in the State include the Bhils (with their subdivisions such as Mankar, Nayak,[5] Patlia, Tarvi etc.), Bhilalas, Barelas, Nihals, Gonds and Korkus.

DISTRIBUTION OF TRIBAL POPULATION

The district and pargana-wise distribution of the primitive tribes in the Indore state is shown in Table 8.3. The distribution of tribal population (Figure 20) in Petlawad, Nisarpur, Sendhwa and Segaon is spectacular. Bhikangaon, Khargone and Kannod also have a considerable tribal concentration. Alampur has no primitive tribes, nor have the central regions of Malwa.

MAP 20.

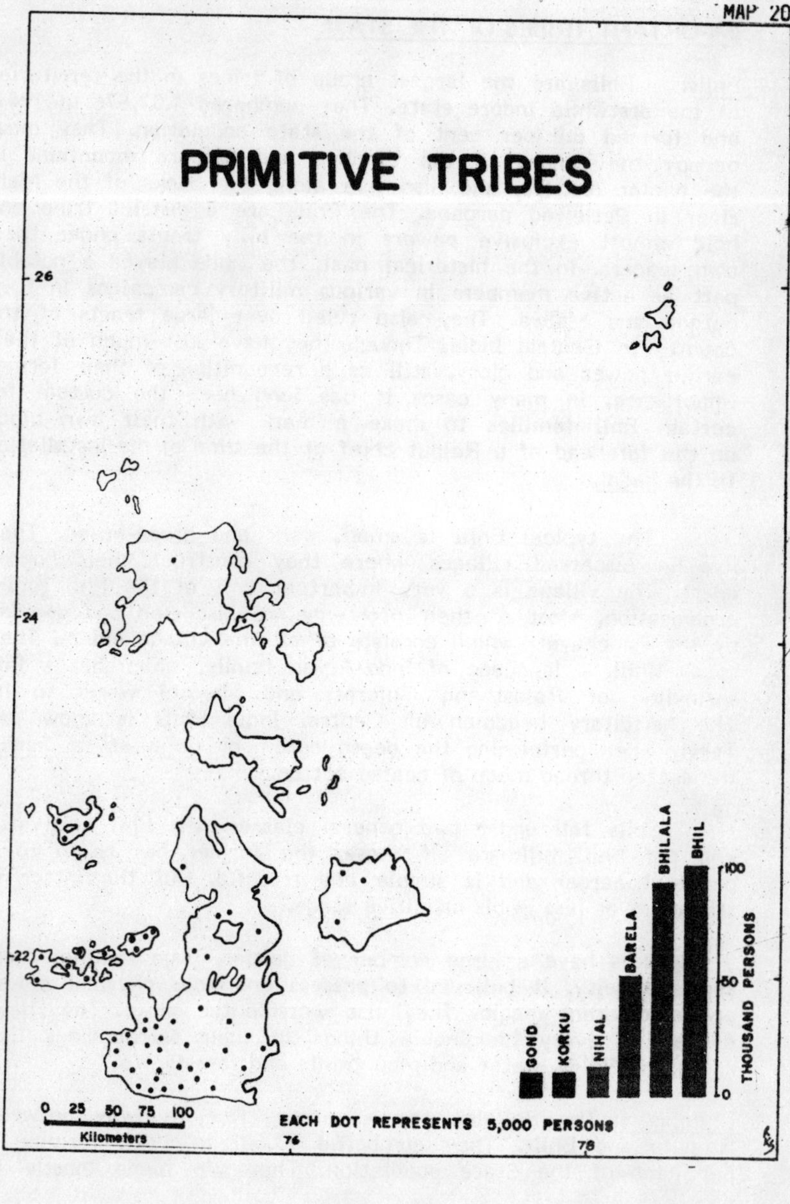
PRIMITIVE TRIBES
26
24
22
0 25 50 75 100
Kilometers
EACH DOT REPRESENTS 5,000 PERSONS
76
78
GOND
KORKU
NIHAL
BARELA
BHILALA
BHIL
0
50
100
THOUSAND PERSONS

The proportion of tribal population is greater just in those areas, where it has not been possible for more civilized peoplees to settle down easily and comfortably.

IMPORTANT TRIBES OF THE STATE

Bhils: Bhils are the largest group of tribes in the territories of the erstwhile Indore state. They numbered 1,02,676 in 1941, and formed 6.8 per cent of the state population. They most occupy the ranges uf the Vindhya and Satpura mountains in the Nimar district, and also live along the banks of the Mahi river in Petlawad pargana. The Bhils are a district tribe and hold almost exclusive powers in the hilly tracts under their own leaders. In the historical past, the Bhils played a notable part as active members in various military campaigns in Rajputana and Malwa. They also ruled over large tracts of the country in Central India. Though they have lost much of their earlier power and glory, still as a recognition of their former importance, in many cases it has long been the custom for certain Bhil families to make a mark with their own blood on the forehead of a Rajput chief at the time of his installation to the gaddi.

The typical Bhill is small, dark and broad-nosed. They live in dispersed villages where they construct their houses apart. The village is a very important unit of the Bhil social organisation. Most of their problems are discussed and decided by the Panchayat, which consists of all the village elders. They speak Bhili, a language of Indo-Aryan family, which has a fair sprinkling of Rajasthani, Gujarati and Marathi words in it. The heriditary headman of Central India Bhils is known as Tarvi, when performing the death ceremony he wears a Janeu, the sacred thread made of coarse cotton.

Bhils fall under two general classes- the Bhil cultivator and the Bhill hillman. Of these, the former has taken to a peaceful career and is simple and truthful. But the latter is still more or less in his primitive stage.

Bhils have a large number of deities. Raja Pantha, their supreme deity, is believed to possess enormous physical power and omnipotent magic. They also worship the snake. They have deities of every imaginable things in their surroundings like crops and fields, water and mountaints and forests etc.[6]

Bhilalas : The Bhilalas are a mixed tribe, midway between Rajputs and Bhils. They numbered 93,682 in 1941, forming 6 per cent of the State population. They are found mostly in

Table 8.3 : Distribution of Primitive Tribes

(Population in 1941)

District/ Pargana	Barela	Bhil	Bhilala	Gond	Korku	Nihal	Total
STATE	53,206	1,02,676	93,682	10,720	10,720	12,182	2,82,602
Indore City	5	1,643	52	50	2	5	1,707
Mhow Cantonment	-	84	4	7	-	1	96
Indore Dist.	-	15,861	223	121	127	-	16,332
Indore	-	2,205	56	34	79	-	2,374
Debalpur	-	3,064	-	2	-	-	3,066
Sawer	-	869	12	-	1	-	882
Mhow	-	9,723	155	85	47	-	10,010
Mahidpur Dist.	-	15,725	3	4	-	-	15,732
Mahidpur	-	462	-	-	-	-	462
Tarana	-	621	2	4	-	-	627
Jirapur	-	334	-	-	-	-	334
Petlawad	-	14,308	1	-	-	-	14,309
Alampur	-	-	-	-	-	-	-
Rampura-Bhanpura Dist.	-	9,118	-	65	-	-	9,183
Nandwai	-	590	-	65	-	-	655
Manasa	-	2,073	-	-	-	-	2,073
Rampura	-	2,735	-	-	-	-	2,735
Bhanpura	-	1,357	-	-	-	-	1,357
Garoth	-	339	-	-	-	-	339
Sunel	-	2,024	-	-	-	-	2,024
Nimar Dist.	53,201	60,182	92,519	1,116	272	11,976	2,19,266
Nisarpur	-	12,138	37,228	-	-	237	49,603
Barwaha	-	3,384	2,091	133	115	1,345	7,068
Maheshwar	1	8,866	809	-	66	35	9,777
Kasrawad	-	1,501	2,578	-	2	1,805	5,966
Segaon	4,008	7,453	28,864	-	2	2,710	43,037
Sendhwa	37,714	6,285	6,584	-	10	515	51,708
Khargone	9,199	5,936	8,964	129	9	3,741	27,978
Bhikangaon	2,279	14,619	5,401	854	68	1,508	24,729
Nemawar Dist.	-	63	881	8,773	10,319	200	20,236
Kannod	-	27	718	4,887	7,619	181	13,432
Khategaon	-	36	163	3,886	2,700	19	6,804

Source : Census of Central India, 1941, Vol.I-Holkar State-Tables, pp. 238-51.

the Nimar district, where they are held in great respect by the Bhil chiefs. Unlike the Bhils, the Bhilalas occupy the more open tracts of the lower Satpuras and the Narmada valley. Two-thirds of them live in the Nisarpur and Segaon parganas of Nimar district, where they form 55 and 42 per cent of the total population. They are mixed with the general population more than any other sections of the Bhils and have greater mixture of Rajput blood. The Bhilalas observe the Hindu rituals and worship Hindu deities.

Some of the Bhilalas held jagirs and one of them was the ruler of a small adjoining State. They occupy the highest position in the social hierarchy among the primitive tribes of the State.

Barelas: Barelas, like Bhilalas, are considered a superior tribe. They numbered 53,206, and are found mostly in Nimar. They are practically confined to the slopes of the Satpuras, especially the higher slopes. In the Holkar state the Barelas were of a recent origin. They were returned for the first time in the 1921 census.

Korkus: The Korkus are an off-shoot of the Munda or the Kolarian tribe. They numbered 10,720 in 1941. They live mainly on the slopes of the Vindhyas in the Nemawar district. They are divided into two groups- Rajkorku and Patharia. The Raj-korkus form the land-owning class and are considered superior to the Patharias in the social scale. They speak a Mundari dialect, which is also called Korku.

Gonds: Though Gonds are the largest tribal group of the country, but in the erstwhile Indore state they had been the smallest tribal community. They numbered 10,136 in 1941 and are confined mainly to the Nemawar district. In Indian chronicles they have been described as a proud people who have earned a name for themselves. From 14th to 19th century, major part of Central India was under the rule of three or four Gond Dynasties. This whole territory was formerly known as Gondwana. In the Scheduled Tribes (Constitution) Order of the Government of India, there are 53 sub-tribes of Gonds, the more important among them being the Marias, Murias, Dhurwas, Agarias and Raj-Gonds, the last being considered to belong to the highest class. Among the Gonds there are several occupational groups such as Agarias, or iron smelters, Ojhas or sooth-sayers, nagarchis or musicians, Solhas or carpenters, Pardhans or minstrels etc. Pardhans are responsible for the preservation of epics, folklore, myths, and legends that have been handed down from word of mouth from generation to generation. The Gonds speak

Gondi language, which belong to the Dravidian family. They worship specific number of deities, ranging from three to seven according to their septs.

Nihals: According to the 1941 census there were 12,182 Nihals, which are confined mostly in Nimar district. Nihals are comparatively better cultivators and occupy the slopes of the Vindhyas and the Satpuras. Some of them are also found in the Nemawar district, particularly in the Kannod pargana.

THE TRIBAL ECONOMY

Ninety nine per cent of the tribal population lives in the villages. There is very little diversification of occupations. Majority of the tribes are agriculturists. They are either cultivators or agricultural labourers. The agricultural labourers also hold small pieces of land and follow cultivation as a major subsidiary occupation.

In the tribal areas the agricultural conditions are totally different from those of open plateau and the plains. In these areas due to undulating nature of the land soil erosion has been extensive. The soils are generally poor in fertility with low moisture-retention quality. Irrigation facilities are non-existant and the tribal cultivators are wholly dependent on rains. Thus in such conditions the tribal farmers get poor returns from their fields. Maize, jowar, cotton and oilseeds (mainly groundnuts) are the principal crops of the tribal areas.

Horticulture presents a good scope in the tribal belt. Growing of fruits like plums, guava, mango, lemon, orange, papaya etc., and of vegetables, especially potatoes have great possibilities. As the tribal areas have vast area under grass, dairy farming can also be very profitably introduced to provide an additional source of income and employment to the tribal people. Poultry and piggery could also be useful programme for the primitive tribs, as they have aptitude, taste and fascination for them.

CULTURAL ATTAINMENTS

Due to their poor economic conditions the primitive tribes in general are culturally backward. They had lesser contact with non-tribal population due to their centuries' old isolation.

For a variety of reasons they remained educationally backward[8]

1. Formal education in tribal areas had a late start, The number of educational institutions in the tribal areas is smaller as compared to other areas. Difficult terrain, lack of communication, sparse population all attribute to the smaller number of schools in tribal areas.

2. Low economic condition of the primitive tribes does not admit of even the smallest expenditure on education. Further to send its grown up boy or girl to school is an economic proposition to a tribal family as it entails dislocation in the traditional pattern of division of labour. The children help the family in many ways; they look after the young ones, do domestic work, help their parents in agricultural operations and collect forest produce.

3. The schools are located at some distance and the young children cannot reach there easily, particularly in rainy season.

4. Lack of interest in academic education, which is not related to tribals' life or surroundings.

5. The number of educational institutions is inadequate.

6. Schools in the tribal areas do not have proper buildings.

7. Teachers in tribal areas come mostly from non-tribal areas. Condition of life and work are difficult for them. In many cases they lack necessary understanding of the people and their work with the required zeal. Indifferent teachers are unable to kindle an urge for education amongst tribal children.

There is a dire need of giving priority to education in the hill areas of the Central India to provide them an honoured place in their fast-growing econoy as 'without education economic development of the tribal people will not be possible.'[9]

The primitive tribes have been and on the whole still are an honest, hospitable, straight-forward and peace loving pegle. As honest people they are frank enough to confess their guilt. They have blind faith in their deities and village elders. They are stay-at-home and self-contained people and believe in their own manners, customs and traditions.

It can easily be surmised that becsause of their backwardness in economy, education, political consciousness, they were easily subjugated by the rulers and could not get themselves free from the feudal dependence for a long time.

SOCIAL FRICTION

Two major communities - Hindus and Muslims, live in complete harmony in the Central India. In the princely days the state used to participate in the festivals of Muslims as s mark of respect to that community. There is popular participation of Hindus in the Muslim festivals, and of the Muslims in the Hindu festivals in Malwa.

But till 1818, when there was domination of Muslim generals in the Holkar Darbar, Muslims had an upper hand in the affairs of the state. After the creation of the erstwhile states of Tonk and Jaora, the Holkar state visualized a peaceful life in which Hindus and Muslims were playing their part without any communal frenzy and trouble.

Though there was no religious friction but caste friction cannot be denied in toto. The Dhangara, the clan of the Holkar dynasty, and the Marathas held inam amd jagir lands, and occupied important civil and military positions in the state as they were kith and kin to the ruler.

Thus the Holkar state never faced any serious trouble due to caste and religious friction. As the ruling family belonged to lower caste of shephards, a pastoral community, caste friction in a state like Indore was beyond imagination.

References

1. B.R.Ambedkar, Thoughts on Linguistic States, Delhi, 1955, p. 34.
2. Census of Central India 1921, Vol. I-Holkar State, Indore, 1921, p. 156.
3. John Malcolm, Memoir of Central India, Vol. I, New Delhi, 1970, p.
4. Surendranath Dube, Census of Central India 1931, Vol. I-
5. ibid., p. 159
6. B.K. Dube and F. Bahadur, A study of the Tribal People and Tribal Areas of Madhya Pradesh, Bhopal, 1967, p. 102.
7. Surendranath Dube, Census of Central India 1931, Vol. I - Holkar State, Indore, 1931, p. 254.
8. Dube and Bahadur, op. cit., pp. 34-35.
9. ibid.,p.47

CHAPTER 9

Summary and Conclusions

The Central India is an area of great physical complexity and immense geopolitical significance. It is an area transitional between the Indo-Gangetic plains and the Deccan, which formed the passageway between the north and south India. Peopled by the primitive tribes like Gonds, Bhils and Santhals with pockets of martial communities like Rajputs and Marathas, who ruled the region prior to independence, the Central India exhibits a unique demographic pattern of its own.

Geopolitically the region formed a great shatterzone, accommodating the great majority of the nearly 600 princely states, most of them now grouped with Madhya Pradesh, Rajasthan and Gujarat. The present day inter-state boundaries in the Central India, with a large number of twists, bends and curves, reflect her merchland position between the powers of the Indo-Gangetic plain and the Deccan.

Malwa, the cultural core of the Central India, underwent a change of masters more often than any other part of the country mainly for her use as the highway to the south. After the disintegration of the Mughal power, the Central India in general and Malwa in particular, came under the sway of the Maratha generals - Holkar, Scindia and Pawar, who established their states at Indore, Gwalior, and Dhar and Dewas. The geography and politics of the Central India have been dominated since then by the actions, reactions and interactions of these Maratha states in the pre-independence days.

Indore state (also called the Holkar state) was a premier state in the Central India, whose territories now form part of the states of Madhya Pradesh and Rajasthan. In two divisions of the erstwhile state-Malwa and Nimar, were divided into five districts and 25 parganas (mahals). Indore was one of the most progressive states in the country, which introduced democratic institutions in the state-craft as early as 1909. Through the local self government institutions-gaothi(village) panchayats, mahal, district and municipal boards, the people of the State participated in the administration of their local affairs.

II

The erstwhile Indore state derived its popular name from its capital city Indore, which appears to be a distortion of Indrapur, the name of the village on which the present city has grown. The state had a very complex vicinal location in the Central Highlands, where it was surrounded by 18 erstwhile princely states of the then Rajputana, Bundelkhand, Bho8al and Central India Agencies.

With an area of 25,729 sq km, the state was of very small size and ranked 15th among the then princely states of the country. The State was extremely incompact and comprised of about 80 discontinued and isolated blocs of territory, scattered about in the Central India, but chiefly situated on the Malwa plateau and in the Narmada valley.

Till 1818, Maheshwar was the capital of the Holkar state. though for a shortwhile Yashwant Rao Holkar directed the affairs of the state from Bhanpura. But his son, Malhar Rao holkar II shifted his capital to Indore in 1818. Indore remained the caaltal of the state till 1948, when after the state's merger in the Indian Union, it became the second capital of Madya Bharat as per Nehru Award.

Indore, as capital city of the Holkar state, was centrally located and was linked by road and rail routes with almost every important town of the state and capital towns of the neighbouring erstwhile princely states.

III

The Holkar state came into existence as an outcome of the conflict between the Mughals and the Marathas in the early 18th century. Malhar Rao Holkar, the celebrated Maratha general, captured a large territory in Malwa after the defeat of the Mughal army at Amjhera (1728). He conquered a number of territories in Central India and Rajasthan, and declared over them the suzerainty of the Peshwa. Malhar Rao was granted by the Peshwa 35 per cent share of the Malwa province alongwith a number of special grants in the Bundelkhand, Rajputana and Deccan in 1730-31, which mark the beginning of the evolution of the Holkar state. The Mughal emperor granted him Sardeshmukhi (a due of 121/2 per cent on the revenue) of Chandore in 1748, the only royal grant to the Holkar family.

After Malhar Rao's death in 1766, his grandson Male Rao succeeded him, but he died after nine months. After his death, his mother Ahilyabai took over the charge of administration. Under her able rule Central India experienced an era of peace and prosperity. She ruled the state till 1795, and was succeeded by Tukoji Rao Holkar, an eminent Maratha general. His death in 1797 marks the beginning of a period of prolonged strife between his sons for securing the gaddi of the Holkars. Finally Yashwant Rao emerged as the head of the Holkar family. He tried to organise the Maratha power and was able to defeat even the combined forces of the Peshwa and Scindia and seized the city of Pune. He was the last maratha general to make his name in the history.

Yashwant Rao's successor, Malhar Rao II, lost most of the Holkar possessions after the Treaty of Mandsaur (1818). His commanders, Amir Khan and Gaffor Khan, were granted the states of Tonk and Jaora respectively under the terms of the Treaty. Malhar Rao's successors-Martand Rao, Hari Rao and Khande Rao, who ruled the state from 1833 to 1844 in succession, proved to be weak and poor administrators, under whom the administration of the state went bad to worse. Tukoji Rao Holkar II was finally installed as head of the Holkar family in 1844. He exchanged a large number of scattered territories to make the Holkar dominions more compact and cohesive. The Holkar State Railway was laid in between 1869-77. Tukoji Rao II breathed his last in 1886, and was succeeded by his eldest son, Shivaji Rao, who abdicted in favour of his son, Tukoji Rao III, in 1903.

Under Tukoji Rao III the state came to be in direct relationship with the Agent to the Governor General in Central India. He too, like his father, abdicted in favour of his minor son, Yashwant Rao Holkar II. The young king was invested with full ruling powers in May 1930. He got the Manpur pargana of the British india transferred to the Holkar state. The state visualised an all-round progress during his regime. He tried to evolve a 'third force' out of states in the wake of India's independence, as he was opposed to the accession of his state in the Indian Union. But later on he signed both the Instrument of Accession and the Standstill Agreement to make his state part of the Indian Union. Maharaja was granted a privy purse of Rs.15 lakhs.

The Holkar state was grouped with other Malwa states into the erstwhile Madhya Bharat, of which Maharaja Yashwant Rao Holkar II was made the Senior Uprajpramukh with a consolidated allowance of Rs.2.5lakhs.

After reorganisation of the states, the former Holkar territories as part of the Madhya Bharat, have become part of Madhya Pradesh. However, Sunel tappa of Bhanpura tahsil was made part of the Rajasthan State.

IV

The state territories lie on the Central Highlands, a compact block of mountains, hills and plateaus, interspersed with valleys and basins. It falls under five physiographic regions- the Malwa plateau, the Vindhya range, the Narmada valley, the Satpura range, and the Chambal-Sindh basin. The Satpura formed its southern boundary. Rivers Chambal, Narmada and their tributaries, and Mahi wash the region and form a radial pattern of drainage. Rivers also defined the state boundaries in all the districts.

The monsoon climate of the region is moderated by the highland nature of its terrain. In cold weather season the average minimum temperatures rarely go below 12.8°C, while in the hot weather season the average ºaximum temperatur s remain below 35°C. There is a marked rainy season from June to September, and the average annual ainfall is about 76 cm. The climatic similarity between the Central India and Maharshtra attracted the Marata generals to capture this region and settle as rulers.

The soils of the state are generally of zonal order. The fertile regur soil of this region produce average yields year after year even without good soil management practices. The leading position of the Indore state among the erstwhile princely states was a direct result of her rich soils- their fertility and productivity.

Forests cover about 26 per cent of the state's former territory and provide a large number of major and minor produces. With proper care the forests can feed a paper and pulp mill. Not very far from the state's old frontiers the Nepanagar Newsprint factory has been established, which gets a lot of raw materials from these forests.

V

The state economy was based essentially on agriculture, which still provides livelihood to most of the people in the Central India. Agriculture supplies raw materials like cotton and oilseeds to the industries besides food for the people and feed for their cattle and other domesticated animals.

The state had about half of its area under cultivation (50 ›er cent) alongwith a large proportion of cultivable waste land 18 per cent). Kharif crops like Jowar, Bajara, maize etc. form ›ulk of the farm produce in the region. The proportion of Kharif :o Rabi in the state had been 73:27. Cotton is the only cash :rop of importance grown in the state territories. Pulses, oil-›eeds and vegetables, and crops like tobacco, and poppy (opium) ›re also produced in these tracts. With better irrigation facili-ties, ample manuring and use of chemical fertilizers, improved seds and implements the per hectare yield can be raised to a large extent.

Minerals have not been important in the state and the 'egional economy. Whatever minerals are found in the area, ›ave not been worked out mainly due to their low ore content. 'races of mangenese, copper and iron ore have been noticed n the state territories. Further search for minerals the area s likely to pay dividends.

The state had fairly well-developed industrial base, which provided subsistence to about one-eighth (12.41 per cent) of her people. Cotton spinning and weaving, dyeing and calico printing, metal utensils, lac bungles, shoe making, hide tanning etc were import cottage industries. There are eight textile mills seven at Indore and one at Sanawad, hundreds of ginning factories, distillery and printing presses located in the area.

Indore has taken a long stride in her industrial development. With planned development of industries, after the independence of the country, the region has experienced a much faster growth of industries, which have made Indore the biggest industrial centre of Central India, in between Bombay and Delhi.

The territories of the erstwhile Indore state have their recognised collecting and distributing centres, which are fed by the weekly markets almost in all the important villages. This Central Indian tract has been an important trading region of the country. Her exports include cotton, cotton cloth, grains, oilseeds, cattle, opium, spices, hides and skins, timber and tobacco whereas imports include piece goods, yarn, salt, sugar, rice, kerosene, machines etc. Most of the import trade includes consumers' goods and exports, the agricultural and allied products.

VI

In the land-locked state of Indore roads and railways provi-ded excellent routes of communication. The state had 1,102 kms of road suitable for motor traffic. Major towns and capitals of erstwhile neighbouring princely states were connected with Indore through these roads. The Government roads (the present--day National Highways), forming part of the inter-state road--network, was supposed to be life-line of the British Government in India, particularly in pre-independence days. The Holkar state had 255 kms of these trunk roads connecting Mhow and Neemuch cantonments with Indore, the seat of the Holkar Government and the Central India Agency.

The railways, both broad and metre gauge, had a route length of 203 kms within the territories of the erstwhile Indore state. While earlier these railways were part of the Holkar State Railway, Bombay Baroda and Central India Railway, and the Great Indian Peninsular Railways, they now form part of the Western Railway.

The roads and the railways of the state had been one of the most important factors exercising a unifying influence for the emotional integration of its people, who lived in a large number of scattered and isolated tracts of territory in the Central India.

As all the military routes to the Deccan, Gujarat and to sea ports on the western coast pass through Malwa, the British Government tried to keep a tight control over the Central Indian tract. The trunk routes passing through the region provided an excellent opportunity to the Britons in maintaining such a grip.

The Indore state had a population of 15,13,966 in 1941, with a density of 55 persons to the sq km. After 1818, the State had registered three-fold increase in its population. However the density of population varied greatly from one district to the other. While Indore and Alampur recorded a density of above 150 persons per sq km, Nemawar had only 35. Population concentration was much more in the fertile and productive areas in comparison to the hilly tracts of the state.

In the Indore state 38 per cent of the population was under 15 years age-group, and only 2,42 per cent above 65 years age-group in 1941. There were 931 females for every 1,000 males. The Holkar state was essentially a Hindu state, where Hindus formed 89.91 per cent of total population. Muslims, Christians, Jains, Sikhs, Parsis, Jews and Buddhists together constituted about one-tenth of her population.

The people speak a number of languages but five of them

are important, i.e. Hindi, Bhili, Marathi, Gujarati and Gondi. They all belong to the Indo-European family of languages. Hindi is spoken by about four-fifths of population and is universally understood throughout the region. Hindi was made the official language of the Holkar state as early as 1901.

The number of literate persons in 1941 was 2,07,681 in the state constituting 14.65 per cent of the total population. The literacy among the males had been 23.32 per cent and among the femaless 5.19 per cent.

By careful planning of the resources, both natural and cultural, the fast-growing population of the region can be provided suitable amenities for a better living.

VIII

In the state the social structure was built up on the caste system, a special feature of the Hindu civilization and culture. Caste, as any where else in India, is based on endogamy, traditional occupation, and a common origin. The castes of the region fall under four categories-functional, secretarian, tribal and mixed. There has been considerable diversity in the caste composition of the state. Some of the castes are depressed and some primitive. Brahmins, Banias, Rajputs and Marathas form advanced group of people in the Indore area.

The depressed classes, also termed as Harijans, belong to that group of the Hindu population which work under certain disabilities attached to certain occupational castes. Untouchability has been regarded as the prime test of declaring a caste depressed or otherwise. But untouchability itself is a relative term. It has never existed in Malwa in its pronounced form.

The state had a large proportion of the tribal population, about one-fifth (18.66) of the total population. The major tribes inhabiting this region include the Bhils, Bhilalas, Barelas, Nihals, Gonds and Korkus. They mostly occupy the hill ranges of the Vindhya and Satpuras in Nimar district, and the Narmada valley in the Nemawar area.

Most of the tribal population earn their bread from agriculture. But their agricultural methods are primitive and need a revolutionary substitution by improved technology and improved implements. For improvement of their social conditions educational facilities have to be provided to every young boy or girl of the school-going age. They need facilities for technical education to develop industrial skill among them for proper utilization of their natural resources, which in turn may provide an economic face-lift to the region.

In the Holkar state there have been almost absence of social friction in the name of caste, creed or religion. Of course there was a little leaning towards the Marathas and Dhangars as they belonged to the clan of the then ruling family of Indore.

Various factors enumerated above had been responsible for the consolidation of the Holkar state as an important politico-geographic unit in the Central India. After merger of the state with the Indian Union in 1948, indore has emerged as a powerful economic, political and administrative base for the state of Madhya Pradesh. With a very rich resource base - both physical and cultural, in the vicinity, Indore is poised to become the focus of the Central India.

Indore, with a population of 8,27,071 in 1981, is not only the biggest city of Madhya Pradesh but also her largest indutrial and commercial centre. It has great possibilities for the development of a variety of industries like dairying, sugar, paper and pulp, aluminium foundry and forge plant, rolling mills, copper sulphate etc. Indore and her suburbs form the largest urban aggglomeration (over one million of population) in the Central India. With her beautiful palaces, spacious parks, industrial culture and economic activities, educational, religious, financial and social institutionss, Indore symbolizes the cultural zenith of Malwa and Central India, which constitute the heartland of the country.

Appendices

APPENDIX I

TREATY OF RAJPUR GHAT (1805)

TREATY WITH JESWUNT RAO HOLKAR WITH THE DECLA RATORY ARTICLE ANNEXED - 1805

TREATY OF PEACE AND AMITY BETWEEN THE BRITISH GOVERNMENT AND JESWUNT RAO HOLKAR

Whereas disagreement has arisen between the British Government and Jeswunt Rao Holkar, and it is now the desire of both parties to restore mutual harmony and concord, the following Articles of Agreement are therefore concluded between Lieutenant-Colonel John Malcolm on the part of the Honourable Company, and Sheikh Hubeeb Oolla and Balla Ram Seit on the part of Jeswunt Rao Holkar, the said Lieutenant-Colonel John Malcolm having special authority for that purpose from the Right Honourable Lord Lake, Commander-in-Chief, & c., His Lordship aforesaid being invested with full powers and authority from the Honourable Sir George Hilaro Barlow, Governor General & c., & c., and the said Sheikh Hubeeb Oolla and Balla Ram Seit also duly invested with full powers on the part of Jeshwunt Rao Holkar.

ARTICLE 1

The British Government engages to abstain from the prosecution of hostilities against Jeswunt Rao Holkar and to consider him henceforward as the friend of the Honourable Company, Jeswunt Rao Holkar agreeing on his part to abstain from all measures and proceedings of a hostile nature against the British Government and its allies, and from all measures and proceedings in any manner directed to the injury of the British Government or its allies.

ARTICLE 2

Jeswunt Rao Holkar hereby renounces all right and title to the districts of Tonk, Rampoora, Boondee, Lekherree, Samay-

dee, Bhamungaum, Dase, and other places north of the Boondee hills, and now in the occupation of the British Government.

ARTICLE 3

The Honourable Company hereby engages to have no concern with the ancient possessions of the Holkar family in Mewar, Malwa, and Harrowtee, or with any of the Rajahs situated to the south of Chumbul; and the Honourable Company agrees to deliver over immediately to Jeswunt Rao Holkar such of the ancient possessions of the Holkar family in the Deccan now in the occupation of the Honourable Company, as are situated south of the river Taptie, with the exception of the fort and pergunnah of Chanderee, the pergunnahs of Amber and Sengham, and the villages and pergunnahs situated to the southward of the river Godavery, which will remain in possession of the Honourable Company. The Honourable Company, however, in consideration of the respectability of the Holkar family, further engages that, in the event of the conduct of Jeswunt Rao Holkar being such as to satisfy the state of his amicable and peaceable intentions towards the British Government and its allies, it will at the expiration of eighteen months from the date of this Treaty restore to the family of Holkar the fort of Chanderee and its districts, the pergunnahs of Amber and Sengham, and the districts formerly belonging to the Holkar family, situated to the south of the Godavery.

ARTICLE 4

Jeswunt Rao Holkar hereby renounces all claims to the district of Koonch in the province of Bundelkhund and all claims of every description in that province; but in the event of the conduct of Jeswunt Rao Holkar being such as to satisfy the British Government of his amicable intentions towards that state and its allies, the Honourable Company agrees at the expiration of two years from the date of this Treaty to give the district of Koonch in Jaghire to Beema Bai, the daughter of Jeswunt Rao Holkar, to be hold on under the Company's Government on the same terms as that now enjoyed by Balla Bai.

ARTICLE 5

Jeswunt Rao Holkar hereby engages never to entertain in his service Europeans of any description, whether British subjects or others, without the consent of the British Government.

ARTICLE 6

Jeswunt Rao Holkar hereby renounces all claims of every description upon the British Government and its allies.

ARTICLE 7

Jeswunt Rao Holkar hereby engages not to admit into his council or service Sarjee Rao Ghautkee, as that individual has been proclaimed an enemy to the British Government.

ARTICLE 8

Upon the foregoing conditions Jeswunt Rao Holkar shall be permitted to return to Hindustan without being molested by the British Government and the British Government will not interfere in any manner in the concerns of Jeswunt Rao Holkar. It is, however, stipulated that Jeswunt Rao Holkar shall immediately, upon the Treaty being signed and ratified, proceed towards Hindustan, by a route which leaves the towns of putteeala, Kythul, Jhind. and the countries of the Honourable Company and the Rajah of Jeypore, on the left; and Jeswunt Rao Holkar engages on his route to make his troops abstain from plunder, and that they shall commit no act of hostility in any of the countries through which they may pass.

ARTICLE 9

This Treaty, consisting of nine Articles being this day settled by Lietuenant-Colonel John Malcolm on the part of the Honourable Company, and by Sheikh Hubeeb Oolla and Balla Ram Seit on the part of Jeswunt Rao Holkar, Lieutenant-Colonel John Malcolm has delivered one copy thereof, in Persian and English signed and sealed by himself, and confirmed by the seal and signature of the Right Honourable Lord Lake, to the said Sheik Hubeeb Oolla and Ball Ram Seit, who, on their part, have delivered to Lieutenant-Colonel John Malcolm a counterpart of the same, signed and sealed by themselves, and engage to deliver another copy thereof, duly ratified by Jeswunt Rao Holkar, to the Right Honourable Lord Lake, in the space of three days, the said Lieutenant-Colonel John Malcolm also engaging to deliver to them a counterpart of the same, duly ratified by the Honourable the Governor General in Council, within the space of one month from this date.

Done in Camp, at Rajpoor Ghaut, on the Banks of the Beas river, this 24th day of December, A.D. 1805, corresponding with the 2nd of Shawal, in the year of the Hegira 1220.

John Malcolm
Sheikh Hubeeb Oolla
Balla Ram Seit

Declaratory article annexed to the treaty of peace and amity concluded between the British Government and Maharajah Jeswunt Rao Holkar through the agency of the right Honourable Lord Lake on the 24th December 1805.

Whereas by the second Article of the above mentioned Treaty Maharaja Jeswunt renounce all right and title to the districts of Tonk, Rampoora, Boondee, Lakherree, Sameydee, Bhaumangaum, Dase, and other places north of the Boondee hills, and now in occupation of the British Government; and whereas it has been understood that the Maharaja attaches great value to the district of Tonk, Rampoora, and other districts in the vicinity, which constituted the ancient possessions of the Holkar family, and the relations of amity and peace being now happily restored between the British Government and Maharaja Jeswunt Rao Holkar, the British Government is desirous of gratifying the wishes of the Maharaja to the utmost practicable extent consistent with the considerations of equity, and of manifesting its solicitude to cultivte the friendship and goodwill of the Maharaja; therefore the British Government hereby agree to consider the provisions of the second Article of the Treaty aforesaid to be void and of no effect, and to relinquish all the districts of Tonk, Rampura, and such other districts in their vicinity as were formerly in the possession of the Holkar family, and are now in the occupation of the British Government.

Done on the River Ganges, the 2nd day of February 1806.

G.H. BARLOW

Source : Aitc[illegible]son, A Collection of Treaties, Engagements and Sanads re[illegible] [illegible]dia and Neihbouring Countries, Vol. IV, Calcutta, 19[illegible] [illegible]28.

APPENDIX II

YASHWANT RAO HOLKAR'S LETTER TO VYANKOJI BHOSLE OF NAGPUR ON 15 FEBRUARY 1806

"The Maratha state had been grasped by foreigners. To resist their aggression, God knows, how during the last two and a half years I sacrificed everything, fighting night and day, without a moment's rest. I paid a visit to Daulat Rao Sindhia and explained to him how necessary it was for all of us to join in averting foreign domination. But Daulat Rao failed me. It was mutual cooperation and goodwill which enabled our ancestors to build up, the Maratha states. But now we have all become self-seekers. You wrote to me that you were coming for my support, but you did not make your promise good. If you had advanced into Bengal as was planned, we could have paralysed the British Government. It is no use of now talking of past things. When I found myself abandoned on all sides, I accepted the offer which the British agents brought to me and concluded the war."

Source: Aitihasik Patra Vyavahar No.394, cited by Sardesai, G.S., New History of the Maratha, Vol. III, Bombay, 1948, p. 437.

APPENDIX III

TREATY OF MANDASOR (1818)

TREATY OF PEACE BETWEEN THE EAST INDIA COMPANY AND MULHAR RAO HOLKAR

Treaty of Peace between the Honourable the East India Company and His Highness Maharajah Mulhar Raw Holkar, his heirs and successors, settled by Brigadier-General Sir John Malcolm, K.C.B., and K.L.S., Political Agent for the Most Noble the Governor General, on the part of the Honourable East India Company, and Tantiah Jog, on the part of His Highness Mulhar Rao Holkar, the said Brigadier-General Sir Lieutenant General Sir Thomas Hislop, Baronet, Commander-in-Chief of the Army of Fort Saint George and the Army in the Deckan, himself invested with full power and authority from the Most Noble Francis Marquis Hastings, K.G., one of His Britannic Majesty's Most Honourable Company to direct and control all the affairs in the East Indias, and the said Tantiah Jog, duly invested with full power on the part of His Highness Mulhar Rao Holkar- 1818.

ARTICLE 1

Peace being established with the Maharajah Mulhar Raw Holkar, the Company's Government agrees that it will not permit any state or any free-booter to be unpunished that shall commit any outrage of hostility against the territories of Maharajah Mulhar Raw Holkar; the Maharajah agreeing on such occasions to lend his utmost assistance by the employment of his troops, or in such other manner as may be requisite, and the British Government will at all times extend the same protection to the territories of Maharajah Mulhar Rao Holkar as to its own.

ARTICLE 2

Maharajah Mulhar Raw Holkar agrees to confirm the engagement which has been made by the British Government with the Nawab Ameer Khan, and to renounce all claims whatever to the territories guaranteed in the said engagement by the British Government to the Nawab Ameer Khan and his heirs.

ARTICLE 3

The pergunnahs of Patchpahar, Dag, Gungrar, Aoor, and others rented by Rajah Jalim Singh, of Kotah, to be ceded in perpetuity to that Chief by the Maharaja Mulhar Raw Holkar, who renounces all claims whatever on these pergunnahs.

ARTICLE 4

Maharajah Mulhar Raw Holkar agrees to cede to the British Government all claims of tribute and revenues of every description which he has or may have had upon Rajpoot princes, such as the Rajahs of Oudepoor, Jeypore, Kotah, Bhoondie, Karawlie, & c.

ARTICLE 5

Mulhar Raw Holkar renounces all rights and titles to any of his territories such as Rampoorah, Basunt, Rajepoorah, Ballocah, Neemseraee, Indeghur, Bhoondie, Lakhearrie, Sanmendah, Bamangaum, Daee and other places within or north of the Bhoondie hills.

ARTICLE 6

Maharajah Mulhar Raw Holkar cedes to the British Government all his territories and claims of every description whatever within and south of the Sutpoorah range of hills, including the fort of Sundwah with a glacis of 2,000 yards, also all his possessions in the province of Candeish and those districts, such as Amber, Ellorah, and others, intermixed with the territories of the Nizam and Paishwah.

ARTICLE 7

In considerations of the cessions made by this Treaty, the British Government bids itself to support a field force to maintain the internal tranquillity of the territories of Mulhar Rao Holkar and to defend them from foreign enemies; this force shall be of such strength as shall be judged adequate to the object. It shall be stationed where the British Government determines tobe best, and the Maharajah Mulhar Raw Holkar agrees to grant some place of security as a depot for its stores.

ARTICLE 8

The Maharajah grants full permission for the purchase of supplies of every description for any British force, acting in the defence of the territories; grain and all other articles of consumption and provisions and all sorts of materials for wearing apparel, together with the necessary number of cattle, horses and camels required for the use of such force, shall be exempted from duties.

ARTICLE 9

Maharajah Mulhar Raw Holkar engages never to commit any act of hostility or aggression against any of the Honourable Company's allies or dependents or against any other power or state whatever. In the event of differences arising, whatever adjustment the Company's Government weighing matters in the scale of truth and justice may determine, shall have the Maharajah entire acquiescence. The Maharajah agrees not to send or receive vakeels from any other state or to have communication with any other states except with the knowledge and consent of the British Resident.

ARTICLE 10

The British Government hereby declares that it has no matter of concern with any of the Maharajah's children, relations, dependents, subjects, or servants, with respect to whom the Maharajah is absolute.

ARTICLE 11

Maharajah Mulhar Raw Holkar agrees to discharge his superflous troops and not to keep a larger force than his revenues will afford. He however agrees to retain in service, ready to cooperate with the British troops, a body of not less than 3,000 horses, for whose regular payment a suitable arrangement must be made.

ARTICLE 12

The Maharajah engages (and the British Government guarantees the engagement) to grant to Nawab Gaffoor Khan his present jaidad of the districts of Sunjeet, Mulhargurh, Tal-Mundawal, Jawrah, Burroda; the tributes of Peeplowdah, with the sayar of the whole. These districts shall descend to his heirs on the condition that the said Nawab and his heirs shall maintain independent of the sebundy for his pergunnahs, and his personal attendants, in constant readyness for service, a body of 600 select horse; and further, that this quota of troops shall be hereafter increased in proportion to the increasing revenue of the districts granted to him.

ARTICLE 13

Mulhar Rao Holkar engages never to entertain in his service Europeans or Americans of any description without the knowledge and consent of the British Government.

ARTICLE 14

In order to maintain and improve the relations of amity and peace hereby established, it is agreed that an accredited Minister from the British Government shall reside with the Maharajah Mulhar Raw Holkar, and that the latter shall be at liberty to send a vakeel to the Most Noble the Governor General.

ARTICLE 15

All the cessions made by this Treaty to the British Government or its allies shall take effect from the date of this Treaty, and the Maharajah relinquishes all claims to arrears from these cessions. The possessions lately conqured by the British Government shall be restored to the Maharajah.

The Parwannahs for the mutual delivery of these cessions shall be issued without delay, and the forts ceded shall be given up with their military stores and in all respects in their present condition.

ARTICLE 16

The English Government engages that it will never permit the Paishwah (Sree Munt) nor any of his heirs and descendents to claim or exercise any soverign rights or power whatever over the Maharaja Mulhar Raw Holkar, his heirs and descendents.

ARTICLE 17

This Treaty consisting of seventeen Articles, has been this day settled by Brigadier-General Sir John Malcolm, acting under the direction of His Excellency Lieutenant-General Sir Thomas Hislop, Baronet, on the part of the Honourable Company, and by Tantiah Jog, on the part of Mulhar Raw Holkar. Sir John Malcolm has delivered one copy thereof in English and Persian, signed and sealed by himself, to the said Tantiah Jog to be forwarded to Maharajah Mulhar Raw Holkar, and has received from the said Tantiah Jog to be forwarded to Maharajah Mulhar Raw Holkar, and has received from the said Tantiah Jog a counterpart of the same Treaty, signed and sealed by him.

Sir John Malcolm engages that a copy of the said Treaty, ratified by the Most Noble the Governor General, in every respect a counterpart of the Treaty now executed by himself, shall be delivered to Tantiah Jog to be forwarded to the Maharajah, within the period of one month, and on the delivery of such copy to the Maharajah, the Treaty executed by Sir John Malcolm, under the immediate direction of His Excellency Sir Thomas Hislop, Baronet, shall be returned; and Tantiah Jog in like manner engages that another copy of the said Treaty, ratified by the Maharajah Mulhar Raw Holkar, in every respect the counterpart of the Treaty now executed by himself, shall be delivered to Sir John Malcolm, to be forwarded to the Most Noble the Governor General, within the space of two days from this date, and on the delivery of such copy to the Most Noble the Governor General, the Treaty executed by Tantiah Jog, by virtue of the full powers and authority vested in him as above mentioned, shall be returned.

Done at Mundessoor this Sixth day of January, A.D. one thousand eight hundred and eighteen, on the twenty-ninth day of Suffer, in the year of the Hejira 1233.

JOHN MALCOLM, Br.-Genl.,
P.A.. Govr.-Genl.

VITUAL PANT TANTIAH JOG

Ratified by His Excellency the Govr.-Genl
in Camp at Oochar, this 16 January 1818. HASTINGS

APPENDIX IV

KHARITA TO MAHARAJA HOLKAR ON EXCHANGE OF TERRITORIES (1881)

KHARITA TO HIS HIGHNESS MAHARAJ ADHIRAJ RAJ RAJESHWAR SAWAL TOOKAJEE RAO HOLKAR BAHADUR, G.C.S.I., C.I.E., OF INDORE-1881

My Honoured and Valued Friend,

The Agent to the Governor General in Central India has at length been able to report to me the conclusion of the negotiations for the exchange of territory, which have been so long pending between Your Highness and the Government of India.

I have now the pleasure to inform Your Highness that I approve and confirm the transfer to the possession of Your Highness of the 231 villages in Taluka Nimawar, and the 176 villages in the district of Nimar, of which the names, revenues, and other particulars are given in schedules A and B of the memorandum by Lieutenant-General Sir H.Daly enclosed in this letter.

The memorandum correctly summarises the results of the negotiations, and I feel assured that Your Highness will not fail to perceive, in the constant desire to meet Your Highness' wishes, which have been evinced by my predecessors thoughout the course of the negotiations, evidence of that friendly spirit which I am confident will ever characteritize the relations of the British Government with Your Highness' loyal and distinguished family.

I beg to express the high consideration I entertain for Your Highness, and to subscribe myself.

Your Highness' sincere
friend,
Viceroy & Governor-
General of India.

Simla:
The 225th May 1881.

MEMORANDUM OF TERRITORIAL EXCHANGES EFFECTED BETWEEN THE GOVERNMENT OF INDIA AND HIS HIGHNESS MAHARAJA HOLKAR,
G.C.S.I., FROM 1861 to 1880.

Whereas the Government of India deemed it advisable to obtain from His Highness Maharaja Holkar, by exchange of territory, the proprietary rights held by His Highness in certain lands in the Deccan, negotiations to effect an exchange of territory were commenced in 1861 by Sir Richmond Shakespear, Agent to the Governor General for Central india.

2. The lands in which His Highness ceded proprietary rights were -

9 Villages of Taluka Chandore.
6 Villages of Poona Collectorate.
3 Villages in Ahmednagar Collectorate.
3 Jaghire villages in Ahmednagar Collectorate.
1 Jaghire village in Sholapur Collectorate.
5 Jaghire villages in the Bulandshahar district of the North-Western Provinces.

3. The lands given in exchange in full sovereignty by the Government of India were-

231 Villages in Taluka Nimawar of which were
Khalsa Villages 134
Quit-rent 55
Rent-free 42
176 villages in the district of Nimar, of which -
113 were villages in pergunnah Barwai.
32 were villages in pergunnah Dhargaon.
30 were villges in pergunnah Kasrawad.
1 Mandlesur.

4. Schedule 'A'[1] attached gives the names of all the villages noted above, together with the revenue as fixed by agreement for each.

5. Schedule 'B'[2] gives the gross and net revenue as finally fixed and agreed to by the contracting parties of the lands given and taken in exchange, as well as the Huqs, cash payments, etc., debitable to each.

6. The following is the detail of the negotiation step by step -

The Satwas-Nimawar pergunnah was transferred to His Highness Maharaja Holkar on 23rd June 1861. No equivalent in land or cash was received until 31st July 1865 on which date the Chandore estate was taken over from Maharaja Holkar.

Reference

1. Aitichison, C.U., A Collection of Treaties, Engagements and Sanads relating to India and Neighbouring Countries, Vol. IV, Calcutta, 1933, pp. 48-49 (not included in this Appendix).
2. Ibid., pp.60-61 (not included in this Appendix).

The Nimar pergunnahs were transferred to the Maharaja on 23rd October 1867, and the Bulandsahar Jaghire taken over from His Highness on 27th October 1867.

The Poona villages in the Deccan were taken over in February 1867, January and April 1968, and March and September 1872.

7. Tbe concessions made to the Maharaja in the course of the Satwas-Chandore exchange were as follows-

I. In exchange for the proprietary right in nine villages in the Deccan His Highness received 231 villages in Satwas-Nimawar in perpetuity.

II. The Government of India accepted the Maharaja's own valuation of Rupees 33,214-15-8 for his nine villages, although this sum was in excess of what he had already agreed to accept, and obviously included several items which the Government of India would be unable to recover.

III. The district of Satwas-Nimawar was valued at the amount of its land revenue only, the abkari revenue being excluded from the calculation. The net gain to the Maharaja on this account was Rupees 2,156-10-6 a year.

IV. The claim to a deduction of Rs.3,209 from the amount of Holkar's Deshmukhi huqs, in commutation of service was waived, and the total amounts of these huqs, Rupees 11,410, was credited to the land instead of to the cash account to which they properly belonged.

V. No interest has been claimed on account of the deferred payment to the Government of India of revenues of Satwas-Nimawar, which, without giving any equivalent, the Maharaja enjoyed from June 1861 to July 1865. The annual land revenue derived from the district by the Government of India before transfer was Rupees 28,872. The revenue for the four years, therefore, would amount to Rupees 12,15,488, and on this sum several years' interest has been forgone.

VI. In August 1872, The Government of India yielded to the Maharaja's request to be credited with the sum of Rupees 5,835-15 on account of repairs to public works alleged to have been made at his expense in Chandore, after the transfer of that estate to the Government of India.

VII. Finally the capitalization of the Maharaja's annual contribution towards the maintenance of the Mahidpur Contingent and the Malwa Bhil Corps, which it was originally intended should be treatedd in close connection with the territorial exchanges, was in 1865 carried out separately at a loss to the British Government of rather more than 51/2 lakhs of rupees.

8. The corresponding concessions made to the Maharaja in the course of the Wabgaon exchange were as follows-

1. In exchange for the priorietary right only in fourteen villages in the Deccan and five villages in the North--Western Provinces, the Maharaja obtained rights of soverignty over 176 villages in Nimar.

II. The value of the Maharaha's villages, on his own computation amounted to Rupees 38,072 a year. The total revenue of the Nimar lands transferred to His Highness was Rupees 47,701.

III. The lapsed Jaghire of Chotta Kasrawad, with a land revenue of Rupees 2,099 a year, was conceded to the Maharaja.

V. No charge was made to His Highness on account of Government buildings at Mandlesur of the value of Rupees 13,225.

VI. No interest has been claimed on account of deferred payment to the Government of india of the revenue of villages which the Maharaja did not give up for some time after he received possession of the Nimar pergunnahs. Moreover, the Nimar pergunnahs were handed over to Maharaja Holkar under a light assessment at the expiry of the settlement, and the re-assessment at the expiry of the settlement, and the re-assessment since made has given the Indore Durbar three or four times the revenue which the Government of India received from the districts and three or four times the value of Holkar's hereditary lands and huqs in the Deccan.

9. As shown in Schedule B, the annual amount due by the Indore Durbar on account of excess of land made over is Rupees 5,285-6-4. This sum will be deducted annually from the sum of Rupees 30,000 paid by the Government of India on account of Kesorai patan (vide page 338, Volume III, Revised Edition of Aitchison's Treaties and Sunnuds).

H.D. DALY
Agent to the Governor General
for Central India

SIMLA,
The 25th September
1880.

Pearcy, C.E., and Associates, World Political Geography, New York, 1957.

Playne, S., Indin States : A Biographical, Historical and Administrative Survey, London, 1921.

Pounds, Norman J.G., Political Geography, New York, 1963.

Prasad, Rajendra, India Divided, Bombay, 1947.

Prescott, J.R.V., Political Geography, London, 1972.

Rahalkar, N.S., and Kapse, V.T., Census of Central India 1921, Vol. I-Holkar State.

Renner, G.T.and Associates, Global Geography, New York, 1944.

APPENDIX V

INSTRUMENT OF ACCESSION (1947)

Whereas, the Indian Independence Act, 1947, provides that as from the fifteenth day of August, 1947, there shall be set up an independent Dominion known as India, and that the Government of India Act, 1935, shall, with such omissions, additions, adaptations and modification as the Governor General may by order specify be applicable to the Dominion of India;

And whereas the Government of India Act, 1935, as so adapted by the Governor General, provides that an Indian State may accede to the Dominion of India by an Instrument of Accession executed by the Ruler thereof:

Now therefore I ______________, Ruler of _____ in the exercise of my sovereignty in and over my said State do hereby execute this, my instrument of Accession, and

1. I hereby declare that I accede to the Dominion of India with the intent that the Governor General of India, the Dominion Legislature, the Federal Court and any other Dominion authority established for the purpose of the Dominion shall, by virtue of this my Instrument of Accession but subject always to the terms thereof, and for the purposes only of the Dominion, exercise in relation to the state of ________(hereinafter referred to as 'this State') such functions as may be vested in them by or under the Government of India Act, 1935, as in force in the Dominion of India on the 15th day of August, 1947 (which Act so in force is hereinafter referred to as 'the Act').

2. I hereby assume the obligation of ensuring that due effect is given to the provisions of the Act within this state so far as they are applicable therein by virtue of this my Instrument of Accession.

4. I hereby declare that I accede to the Dominion of India on the assurance that if an agreement is made between the Governor General and the Ruler of this state whereby any functions in relation to the administration in this state of any law of the Dominion legislature shall be exercised by the Ruler of this state, then any such agreement shall be deemed to form part of this Instrument and shall be construed and have effect accordingly.

5. The terms of this my Instrument of Accession shall not be varied by any amendment of the Act or of the Indian Independence Act, 1947, unless such amendment is accepted by me an Instrument supplementary to this Instrument.

6. Nothing in this Instrument shall empower the Dominion Legislature to make any law for this state authorising the compulsory acquisition of land for any purpose, but I hereby undertake that should the Dominion for the purposes of a Dominion law, which applies in this state, deem it necessary to acquire any land, I will at their request acquire the land at their expense or if the land belongs to me transfer it to them on such terms as may be agreed, or in default of agreement, determined by an arbitrator to be appointed by the Chief Justuce of India.

7. Nothing in this Instrument shall be deemed to commit me in any way to acceptance of any future Construction of India or to fetter my discretion to enter into agreements with the Government of India under any such future Constitution.

8. Nothing in this instrument affects the continuance ot my sovereignty in and over this state, or save as provided by or under this Instrument, the exercise of any powers, authority and rights now enjoyed by me as Ruler of this state of the validity of any law at present in force in the state.

9. I hereby declare that I execute this Instrument on behalf of this state and that any reference in this Instrument to me or to the Ruler of the state is to be construed as including a reference to my heirs and successors.

Given under my hand this ______________________ ______ day of August, Nineteen hundred and forty-seven.

____

I do hereby accept this Instrument of Accession. Dated this ______________________ day of August, Nineteen hunded and forty-seven.

(Governor General of India)

Source : Government of India, Ministry of States, White Paper on Indian States, Delhi, 1950, pp.165-66.

APENDIX VI

THE COVENANT FOR THE FORMATION OF MADHYA BHARAT

(22.4.1948)

We, the Rulers of Gwalior, Indore and certain other states in Central India,

Being convinced that the welfare of the people of this region can best be secured by the establishment of a state comprising the territories of our respective states, with a common Executive, Legislature and Judiciary;

And having resolved to entrust to a Constituent Assembly of elected representatives of the people the drawing up of a democratic Constitution for the state within the framework of the Constitution of India, to which we have already acceded. and to this Covenant;

Do hereby, with the concurrence and guarantee of the Government of India, do enter into the following Covenant.

ARTICLE I

In this Covenant -

(a) 'Covenanting State' means any of the states mentioned in Schedule I, the Ruler of which has, whether by himself or by a duly authorised representative, signed this Covenant;

(b)'Covenanting Major State' means the state of Gwalior or the State of Indore;

(c) 'Convenanting Major State' means any Covenanting State which is mentioned in Part A of Schedule I:

(d) 'Covenanting Non-Salute State' means any Covenanting state which is mentioned in Part B of Schedule I; and

(e) unless there is any thing repugnant in the subject or context, reference to the Ruler of a state include any person or persons for the time being exercising the powers of the Ruler, whether by reason of his minority or for any other reason.

ARTICLE II

(1) The Covenanting states agree-

(a) To unite and integrate their territories in one state with a common executive, Legislature and judiciary, by the name of the United state of Gwalior, Indore and Malwa (Madhya Bharat), hereinafter referred to as 'the United state'; and

(b) To incude in the United state any other state, the Ruler of which agrees, with the approval of the Government of India, to the merger of his state in the United state.

(2) The terms of all the agreements of merger referred to in clause (b) of paragraph 1 of this Article shall be binding on the United state and shall be deemed to be part of this Covenant.

ARTICLE III

(1) There shall be a Council of Rulers consisting of the Rulers of the Covenanting Salute States, the Ruler of Kurwai and one Ruler to be elected from among themselves by the rulers of the Covenanting Non-States other than Kurwai;

Provided that no Ruler who is less than 21 years of age shall be a Member of the Council.

(2) The Council of Rulers shall elect at a meeting, one member of the Council to be the President, another to be the Senior Vice president, and two others to be Junior Vice Presidents of the Council, and the President so elected shall be the Raj Pramukh of the United State;

Provided that the Rulers of the Covenanting Major states shall not take part in the voting for the election of either Junior Vice Presidents.

(3) For the purpose of the elections referred to in the preceding paragraph, every member shall have such number of notes as is equal to the number of lakhs in the population of his state as ascertained at the last preceding census (any fraction more than half a lakh being reckoned as one lakh and any other

fraction being ignored), provided that every member shall have atleast one vote.

ARTICLE III

(1) There shall be a Council of Rulers consisting of the Rulers of the Covenanting Salute States, the Ruler of Kurwai and one Ruler to be elected from among themselves by the rulers of the Covenanting Non-States other than Kurwai;

Provided that no Ruler who is less than 21 years of age shall be a Member of the Council.

(2) The Council of Rulers shall elect at a meeting, one member of the Council to be the President, another to be the Senior Vice president, and two others to be Junior Vice Presidents of the Council, and the President so elected shall be the Raj Pramukh of the United State;

Provided that the Rulers of the Covenanting Major states shall not take part in the voting for the election of either Junior Vice Presidents.

(3) For the purpose of the elections referred to in the preceding paragraph, every member shall have such number of notes as is equal to the number of lakhs in the population of his state as ascertained at the last preceding census (any fraction more than half a lakh being reckoned as one lakh and any other fraction being ignored), provided that every member shall have atleast one vote.

(4) A Ruler elected as the President or as a Vice President of the Council shall be entitled to hold office as such-President or Vice President, as the case may be for a term of five years from the date on which he enters upon the duties of that office.

(5) Notwithstanding anything contained in the preceding paragraphs of this Article, the present Rulers of Gwalior, Indore, Dhar and Khilchipur shall respectively be the first President, Senior Vice President and Junior Vice Presidents of the Council o Rulers, and shall enter upon the duties of their respective offices on the 11th day of May 1948. The said President and Senior Vice President shall be entitled to hold office during their life-time, and the said Junior Vice Presidents shall be entitled to hold office for a term of five years from the said date.

ARTICLE IV

(1) There shall be paid to the Raj Pramukh from the revenue of the United State a sum of Rs.2,50,000 per year as consolidated allowance in order that he may be enabled to discharge conveniantly and with dignity the duties of his office.

(2) There shall in each year be paid to the Senior Vice President from the revenue of the United State a sum of Rs.2,50,000 as consolidated allowance in order to enable him to discharge conveniently and with dignity the duties of his office.

(3) There shall be paid from time to time to each Junior Vice President such allowance as the Raj Pramukh may consider appropriate to cover expenses in travelling in the discharge of such official duties as he may be directed to perform by the Raj Pramukh.

(4) If the Raj Pramukh is, by reason of absence or illness or for any other reason, unable to perform the duties of his office, those duties shall until he has resumed them, be performed by the Senior Vice President. During such period the Senior Vice President shall be entitled to the same salary, allowances and other amenities as the Raj Pramukh.

ARTICLE V

(1) There shall be a Council of Ministers to aid and advise the Raj Pramukh in the exercise of his functions except those under Article VII.

(2) The Ministers shall be chosen by, and shall hold office during the pleasure of the Raj Pramukh.

ARTICLE VI

(1) The Ruler of each Covenanting State shall, as soon as may be practicable, and in any event not later than the first day of July 1948, make over the administration of his State to the Raj Pramukh; and thereupon-

(a) all rights, authority and jurisdiction belonging to the Ruler which appertain, or are incidental, to the Government State shall vest in the United State and shall hereafter be exercisable only as provided by this Covenant or by the Constitution to be framed thereunder;

(b) all duties and obligations of the Ruler pertaining or

incidental to the Government of the Covenanting State shall devolve on the United State and shall be discharged by it;

(c) all the assets the liabilities of the Convenanting State shall be the assets and liabilities of the United State; and

(d) the military forces if any, of the Covenanting State shall become the military forces of the United State.

(2) When in persuance of any such agreement of merger as is referred to in clause (b) of paragraph (1) of Article II, the administration of any other State is made over to the Raj Pramukh, the provisions of clauses (a), (b), (c) and (d) of paragraph (1) of this Article shallapply in relaation to a Covenanting State.

ARTICLE VII

(1) In this Article, 'scheduled areas' means any of the areas specified in Scheduled II:

Provided that the Raj Pramukh may, in consultation with the Government of India, by proclamation direct that the whole or any specified part of a scheduled area shall cease to be a scheduled area or a part of such an area.

(2) Subject to any directions or instructions that may from time to time be given by the Government of india in this behalf, the authority-

(a) to make laws for the peace and good government of any scheduled area,

(b) to raise, maintain and administer the military forces of the United State, and

(c) to control the administration of the fund in Gwalior known as the Gangajali Fund and of any other existing fund of a similar character in any other Covenanting State,

shall vest exclusively in the Raj Pramukh.

(3) Until other provision is made by an Act of the Legislature of the United State, the right to resume Jagirs or to recognise the succession, according to law and custom, to the rights and titles of a Jagirdar shall vest exclusively in the Raj Pramukh.

(4) Nothing in the preceding paragraphs of this Article shall be deemed to prevent the Raj Pramukh from consulting the Council of Ministers in regard to any of the matters mentioned therein.

ARTICLE VIII

The Raj Pramukh shall, as soon as practicable, and in any event not later than the fifteenth day of June 1948 execute on behalf of the United State an Instrument of Accession in accordance with the provisions of section 6 of the Government of India Act, 1935, and in place of the Instruments of Accession of the several Covenanting States; and he shall be such Instrument accept as matters with respect to which the Dominion Legislature may make laws for the United State all the matters mentioned in List I and List III of the Seventh Schedules to the said Act, except the entries in List I relating to any tax or duty.

ARTICLE IX

Subject to the provisions of this Covenant and of the Constitution to be framed thereunder the executive authority of the United State shall be exercised by the Raj Pramukh either directly or through officers subordinate to him; but the Raj Pramukh may from time to time consult the Senior Vice President, an important matters connected with the administration of the United State. Nothing in this Article shall prevent any competent Legislature of the United State from conferring functions upon subordinate authorities or be deemed to transfer to the Raj Pramukh any functions conferred by any existing law on any court, Judge, officer or local or other authority in a Covenanting State.

ARTICLE X

(1) There shall be formed, as soon as may be practicable, a Constituent Assembly in the manner indicated in Schedule III; and it shall be the duty of that Assembly to frame a Constitution of a unitary type for the United State within the framework of this Covenant and the Constitution of India, and providing for a Government responsible to the Legislature.

(2) The Raj Pramukh shall constitute not later than the first day of August 1948 an interim Legislative Assembly for the United State in the manner indicated in Scheduled IV.

(3) Upon the formation of the Constituent Assembly referred to in paragraph (1) of this Article, the interim Legislative Assembly shall automatically be dissolved, and the Legislative authority of the United State shall vest in the Constituent Assembly.

Provided that until a Constitution framed by the Constituent Assembly comes into operation after receiving the assent of the Raj pramukh, the Raj pramukh shall have powers to make and promulgate Ordinances for the peace and good government of the United State or may part of peace and good government of the United State or may part thereof, and any Ordinance so made shall, for the space of not more than six months from its promulgation have the like force of law as an Act passed by the interim Legislative Assembly, or as the case may be, the Constituent Assembly; but any such Ordinance may be controlled or superceded by any such Act.

ARTICLE XI

(1) The Ruler of each Covenanting State shall be entitled to receive annualy from the revenues of the United State for his privy purse the amount specified against that Covenanting State in Schedule I:

Provided that the sums specified in the Schedule in respect of the Rulers of Gwalior and Indore shall be payable only to the present Rulers of these States and not to their successors for whom provison will be made subsequently.

(2) The said amount is intendedd to cover all the expenses of the Ruler and his family including expenses of his residence, marriages and other ceremonies, etc., and shall subject to the provisions of paragraph (1) neither be increased for any reason whatsoever.

(3) The Raj Pramukh shall cause the said amount to be paid to the Ruler in four equal instalments at the beginning of each quarter in advance.

(4) The said amount shall be free of all taxes, whether imposed by the Government of the United State or by the Government of India.

ARTICLE XII

(1) The Ruler of each Covenanting State shall be entitled to the full ownership, use and enjoyment of all private properties (as distinct from State properties) belonging to him on the

date of his making over the administration of that State to the Raj Pramukh.

(2) He shall furnish to the Raj Pramukh before the first day of August 1948 an inventory of all immovable properties, securities and cash balances held by him as such private property.

(3) If any dispute arises as to whether any item of property is the private property of the Ruler or State property, it shall be referred to such person as the Government of India may nominate in consultation with the Raj Pramukh, and the decison of that person shall be final and binding on all parties concerned:

Provided that no such dispute shall be so referable after the first of July 1949.

ARTICLE XIII

The Ruler of each Covenanting State, as also the members of his family, shall be entitled to all the personal privileges, dignities and titles enjoyed by them, whether within or outside the territories of the State, immediately before the 15th day of August 1947.

ARTICLE XIV

(1) The succession, according to law and custom to the gaddi of each Covenanting State, and to the personal rights, priviliges, dignities and titles of the Ruler thereof, is hereby guaranteed.

(2) Every question of disputed succession in regard to Covenanting State shall be decided by the Council of Rulers after referring it to a bench consisting of all the available judges of the High Court of the United State and in accordance with the opinion given by that High Court.

ARTICLE XV

No enquiry shall be made by or under the authority of the United State, and no proceedings shall lie in any court in the United State, against the Ruler of any Covenanting State, whether in a prsonal capacity or otherwise, in respect of anything done or omitted to be done by him or under his authority done or omitted during the period of his administration of that State.

ARTICLE XVI

(1) The United State hereby guarantee either the continuance in service of the permanent members of the public services of each of the Covenanting States on conditions which will be not less advantageous than those on which they were serving on the 15th April 1948 or the payment of reasonable compensation.

(2) The United State further guarantee the continuance of pensions and leave salaries, sanctioned by competent authorities in any of the Covenanting States to members of the public services in that State who have retired, or proceeded on leave preparatory to retirement, and the compassionate allowances granted to dependents of deceased members of those services, before the date on which the administration of that State is handed over to the Raj Pramukh.

(3) The provisions of paragraphs (1) and (2) of this Article shall apply also in relation to the public services of any other State merging in the United States.

ARTICLE XVII

Except with the previous sanction of the Raj pramukh, no proceedings-civil or criminal, shall be instituted against any person in respect of any act done or purporting to be done in the execution of his duty as a servant of any Covenanting State before the date on which the administration thereof is made over to the Raj Pramukh.

ARTICLE XVIII

Notwithstanding anything contained in the preceding provisions of this Covenant, the Rulers of Gwalior and Indore shall continue to have and exercise their present power of suspension, remission or commutation of death sentences in respect of any person who may have been, or is hereafter, sentenced to death for a capital offence committed within the territories of Gwalior or of Indore, as the case may be.

In confirmationm of the above Covenant we append our signatures, on behalf of ourselves, our heirs and successors.

1. Maharaja of Alirajpur 2. Rana of Barwani
3. Maharaj of Dewas(Junior) 4. Maharaj of Dewas (Senior)
5. Maharaja of Dhar 6. Maharaja of Gwalior
7. Maharaha of Indore 8. Nawab of Jaora

9. Raja of Jhabua
10. Raja of Khilchipur
11. Raja of Narsingarh
12. Raja of Rajgarh
13. Maharaja of Ratlam
14. Raja of Sailana
15. Raja of Sitamau
16. Rana of Jobat
17. Thakur of Kathiwara
18. Nawab of Kurwai
19. Rana of Mathwar
20. Rawat of Piploda

The Government of India hereby concur in the above Covenant and guarantee all its provisions. In confirmation whereof Mr. Vapal Pangunni Menon, Secretary to be Government of India in the Minsitry of States, appends his signature on behalf and with the authority of the Government of India.

SECRETARY TO THE GOVERNMENT OF INDIA
MINISTRY OF STATES.

APPENDIX VII

NOTIFICATION REGARDING HANDING OVER ADMINISTRATION OF INDORE STATE TO THE RAJPRAMUKH OF MADHYA BHARAT

(1948)

GOVERNMENT OF HIS HIGHNESS THE MAHARAJA HOLKAR
PRIME MINISTER'S OFFICE

NOTIFICATION

No.38 Dated the 16th June, 1948.

The following is published for general information.

1. Whereas under Article VI of the Covenant entered into by the Rulers of Gwalior, Indore and certain other States in Central India, the Ruler of each covenanting State is required as soon as practicable, and in any event not later than the first day of July 1948, to make over the administration of his State to the Raj pramukh.

2. And whereas the Minister-in-Attendance to His Highness the Maharaja Holkar of Indore in his letter No.236 dated June 9, 1948 has communicated to the Council of Ministers of Indore authority on behalf of his Highness the Maharaja Holkar of Indore to hand over the administration of the Indore State to the Raj Pramukh on June 16, 1948.

3. The undersigned as Prime Minister of Indore, acting on behalf of the Council of Ministers, has accordingly handed over the administration of the Indore State to His Highness the Raj Pramukh, Madhya Bharat Sanyukt Rajya this day of forenoon of June 16, 1948.

V.S. KHODE
PRIME MINISTER

Source: Holkar Government Gazette, No.38 dated 21 June 1948.

APPENDIX VIII

REVISED INSTRUMENT OF ACCESSION (19TH JULY 1948)

Whereas by Instruments of Accession executed in August, November and December, 1947, and February 1948, the Rulers of the States specifiedd in the Schedule heretc have acceded to the Dominion of India ;

And whereas by Article VIII of the said Covenant it is provided that the Raj Pramukh of the United State shall, as soon as practicable, execute on behalf of the United State a fresh Instrument of Accession in accordance with the provisions of Section 6 of the Government of India Act, 1935, accepting as matters with respect to which the Dominion legislature may make laws for the United State all matters mentioned in List I and III of the Seventh Schedule to the said Act, except matters relating to taxation;

Now, therefore, I, Maharajadhiraja Sir George Jivaji Rao Scindia Bahadur of Gwalior, Raj Pramukh of the United State of Gwalior, Indore and Malwa (Madhya Bharat) do hereby execute this Instrument of Accession for and on behalf of the United State and

1. I hereby declare that I accede to the Dominion of India with the intent that the Governor General of India, the Dominion Legislature, the Federal Court and any other Dominion authority, established for the purpose of the Dominion shall, by virtue of this Instrument of Accession but subject always to the terms thereof, and for the purposes only of the Dominion, exercise in relation to the United Staate such functions as may be vested in them by or under the Government of India Act, 1935, as for the time being in force in the Dominion of India (which Act as so in force is hereinafter referred to as 'the Act').

I hereby assume the obligation of ensuring that due effect is given to the provisions of the Act within the United State so far as they are applicable therein by virtue of this Instrument of Accession.

3. I accept all matters enumerated in List I and III of the

Seventh Schedule to the Act as matters in respect of which the Dominion Legislature may make laws for the United State:

Provided that nothing contained in the said Lists or in any other provision of the Act shall be deemed to empower the Dominion Legislature to impose any tax or duty in the territories of the United State or to prohibit the imposition of any duty or tax by the Legislature of the United State in the said territories.

Provided further that where a law of the United State with respect of one of the matters enumerated in the said list III contains any provision repungnant to the provisions of an earlier Dominion law or an existing law with respect to that matter, then, if the law of the United State having been reserved for the consideration of the Governor General the law of the United State shall prevail in the United State but nevertheless the Dominion Legislature may any time enact further legislation with respect to the same matter;

Provided further that no Bill or amendment for making any provision repugnant to any law of the United State, which, having been so reserved, has received the assent of the Governor General, shall be introduced or moved in the Dominion Legislature without the previous sanction of the Governor General.

4. I do hereby declare that I accede to the Dominion of India on the assurance that if an agreement is made between the Governor General and the Raj Pramukh of the United State whereby any functions in relation to the administration in the United State of any law of the Dominion Legislature shall be deemed to form part of this Instrument and shall be construed and have effect accordingly.

5. I further declare that the provisions contained in Part VI of the Act with respect to interference with water supplies shall apply in relation to the United State.

6. The terms of this Intrument of Accession shall not be varied by any of amendment of the Act or of the Indian Independence Act, 1947, unless such amendment is accepted by the Raj Pramukh of the United State by an Instrument supplementary to this Instrument.

7. Nothing in this Instrument shall empower the Dominion Legislature to make any law for the United States authorising the compulsory acquisition of land for any purpose, but should the Dominion for the purposes of a Dominion law which applies

in the United State shall necessary the request into acquire any land, the Raj Pramukh of the United States shall be at the request and at the expense of the Dominion Government acquire the land or if the land belongs to the United State transfer it to them on such terms as may be agreed, or, in default of agreement, determined by an arbitrator to be appointed by the Chief Justice of India.

8. Nothing in this Instrument shall be deemed to commit the United State in any way to acceptance of any future Constitution of India or to fetter the descretion of the Government of the United State to enter into arrangements with the Government of India under any such future constitution.

9. Save as provided by or under this Instrument, nothing contained in this Instrument shall affect the exercise of any powers, authority and rights enjoyed by the Raj Pramukh or the validity of any law for the time being in force in the United State or any part thereof.

10. The Instrument of-Accession executed in August, November and December, 1947 and February 1948, by the Rulers of the States specified in the Schedule hereto are hereby cancelled.

Given under my hand this 19th day of July, 1948.

(sd.) G.J.R. SCINDIA

I do hereby accept this Instrument of Accession.
Dated this 13th day of September. nineteen hundred and forty eight.

(sd.) C. RAJAGOPALACHARI
Governor General of India

SCHEDULE

1. Alirajpur
2. Barwani
3. Dewas (Junior)
4. Dewas (Senior)
5. Dhar
6. Gwalior
7. Indore
8. Jaora
9. Jhabua
10. Khilchipur
11. Narsingarh
12. Rajfarh
13. Ratlam
14. Sailana
15. Sitamau
16. Jobat
17. Kathiwara
18. Kurwai
19. Mathwar
20. Piploda
21. Nimkhera
22. Pathari
23. Muhammadgarh

Bibliography

Aitchison, C.U., A Collection of Treaties, Engagements and Sanads Relating to India and Neighbouring Countries, Calcutta, 1933.

Alexander, Lewis M., World Political Patterns, London, 1957.

Ambedkar, B.R.. Thoughts on Linguistic States, Delhi, 1955.

Barton, W., The Princes of India, London, 1934.

Bhandari, S.R., Bharat Ke Deshi Rajya (Hindi), Indore, 1927.

Carlson, L., Geography and World Politics, Dehradun, 1971.

Chudgar, R., Indian Princes under British Protection, London, 1929.

Cohen, S.B., Geography and Politics in a Divided World, London, 1964.

Dhariwal, L.C., The Indore State Gazetteer, Indore, 1931.

Dhar, S.N., The Indore City and its Vicinity, Allahabad, 1936.

Dube B.K. and Bahadur, F., A Study of Tribal People and Tribal Area of Madhya Pradesh, Bhopal, 1967.

Dube, Surendranath, Census of Central Indian 1941, Vo. I-Holkar State-Tables, Indore, 1941.

Dubey, R.N., Economic Geography of India, Allahabad, 1968.

East, W.G. and Moodie, A.E., The Changing World Studies in Political Geography, New York, 1956.

East, W.G. and Spate, O.H.K., The Changing Map of Asia, New York, 1961.

Eckford, C.E., Holkar State Census Report, 1901

Edwards, S. and Garrett, H.L.O., Mughal Rule in India, Delhi, 1956.

Finch, V.C. Trewartha, G.T., Robinson, A.H. and Hammond, F.H., Elements of Geography - Physical and Cultural, New York, 1957.

Fordham, Paul, The Geography of African Affairs, Middlesex, 1965.

Gandhi, M.K., The Indian States Problem, Ahmedabad, 1941.

Ghurye, G.S., Caste and Races in India, New York, 1946.

Goblet, Y.M., Political Geography and the World Map, London, 1956.

Gottman, J., "Geography and International Relations", World Politics, Vol. 3., 1951.

Gottman, J., "The Political Partitioning of Our World: An Attempt at Analysis, World Politics, Vol. 4, 1952.

Government of India, Ministry of States, White Paper on Indian States, Delhi, 1950.

Government of India, The State Reorganisation Act 1956, Delhi, 1956.

Hawks, E., The Romance of Transport, London, 1936.

Hoard, Final Report on the Settlement of the Indore State, Bombay, 1907.

Holderness, T.W., Peoples and Problems of India, New York, 1912.

Hutton, J.H., Caste in India, New York, 1946.

Hutton, J.H.. Census of India 1931, Vol. I

Jackson, W.A.D., "Whither Political Geography", Annals of the Association of American Geographers, Vol. 48, 1958.

Jones. C.F. and Darkenwald, G.G., Economic Geography, New York, 1966.

Law, B.C., Historical Geography of Ancient India, Paris, 1954.

Luard, C.E. and Dube, R.P., Indore State Gazetteer, Calcutta, 190.

Luard, C. E.,and Dube, R.P., Indore State Gazetteer, Calcutta, 1908.

Luard. C.E.,Census of India 1911, Central India Agency, Calcutta, 1911.

Lumby, F.W.R., The Transfer of Power in India, 1954.

Mac Munn, G., Indian States and Princes, London, 1936.

Malav Itihas Mandal, The Mandlik Papers and the Family, Indore, 1946.

Malgonkar, M., Puars of Dewas Senior, Bombay, 1963.

Majumdar, D.N.,Races and Cultures in India, Lucknow, 1944.

Malcolm, John, Memoir of Central India including Malwa and Adjoining Provinces, New Delhi, 1970.

Martin, J.T.,Census of India 1921, Vol. I Calcutta, 1921.

Menon, V.P., The Story of the Integration of the Indian States, Calcutta, 1956.

Menon, V.P., Transfer of Power in India, New Delhi, 1968.

Moodie, A.E., Geography Behind Politics, london, 1963.

Mukerjee, R.K., Economic Problems of Modern India, London, 1941.

Nag, D.S., The Tribal Economy, Delhi, 1958.

Naik, T.B., The Bhils, Delhi, 1956.

National Council of Applied Economic Research, Techno-Economic Survey of Madhya Pradesh, New Delhi, 1960.

Panikkar, K.M., Geographicl Factors in Indian History, Bombay, 1955.

Panikkar, K.M., Indian States, London, 1944.

Pearcy, C.E., and Associates, World Political Geography, New York, 1957.

Playne, S., Indin States : A Biographical, Historical and Administrative Survey, London, 1921.

Pounds, Norman J.G., Political Geography, New York, 1963.

Prasad, Rajendra, India Divided, Bombay, 1947.

Prescott, J.R.V., Political Geography, London, 1972.

Rahalkar, N.S., and Kapse, V.T., Census of Central India 1921, Vol. I-Holkar State.

Renner, G.T. and Associates, Global Geography, New York, 1944.

Index